Our Green

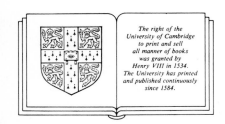

The right of the
University of Cambridge
to print and sell
all manner of books
was granted by
Henry VIII in 1534.
The University has printed
and published continuously
since 1584.

CAMBRIDGE UNIVERSITY PRESS
*Cambridge, London, New York, New Rochelle,
Melbourne, Sydney*

Special Photography by KJELL B. SANDVED *and* EDWARD S. AYENSU

and Living World
The Wisdom to Save It

By EDWARD S. AYENSU, VERNON H. HEYWOOD, GRENVILLE L. LUCAS
and ROBERT A. DEFILIPPS

A Welcome by S. DILLON RIPLEY
An Introduction by HRH THE PRINCE PHILIP
An Epilogue by MRS. INDIRA GANDHI

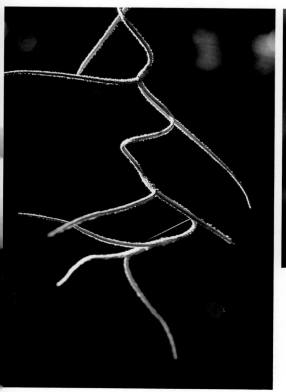

Copublished in 1984 by
Smithsonian Institution Press,
Washington, D.C. and by the Press
Syndicate of the University of
Cambridge, The Pitt Building,
Trumpington Street, Cambridge CB2
1RP, 32 East 57th Street, New York,
New York 10022, U.S.A., 296
Beaconsfield Parada, Middle Park,
Melbourne 3206, Australia

Printed and bound in Great Britain
by Balding + Mansell, Wisbech,
Cambridgeshire

First Edition 5   4   3   2   1

**Library of Congress Cataloging in
Publication Data**
Main entry under title:
Our green and living world.

    Bibliography: p.
    Includes index.
    Supt. of Docs. no.: SI 1.2:W89/3
    1. Nature conservation.
2. Ecology.   3. Conservation of
natural resources.   I. Ayensu,
Edward S.
QH75.093 1984   333.95'16
84-600181
ISBN 0-89599-016-4
ISBN 0-521-26842-7 (Cambridge)

**British Library Cataloguing in
Publication Data**
Our Green and living world.
1. Environmental protection
I. Ayensu, E.S.
333.7'2   TD170

ISBN 0-521-26842-7

Contents

Glimpses of
Paradise

S. Dillon Ripley

WITH FAMILIES OUT FOR A DAY AT THE MUSEUMS, the grass and trees green and flowers in bloom, glimpses of something very much like paradise can be caught right on the Mall in Washington, D.C. I think that the music from the carousel must help, and the heavenly whiff of hot popcorn from the red wagon.

Each one of us treasures our own visions of paradise, but the word itself came to us from the Old Persian *pairidaeza* originally meaning a walled garden. The elegant compounds built by Cyrus and Darius as well as Babylon's famed Hanging Gardens are gone. Yet the basic layout survives. Instead of the graven stone of antiquity, today we find high walls of mud or brick, an enclosure to bar the sun and to hold in puffs of moisture evaporated from fountain or pool. Fruit trees come into flower. Pansies, narcissus, petunias, zinnias, and roses bloom in a riot of color, each in its season.

Such are the perfect little Persian gardens where the wispy grass is never cut, growing to delicious knee-depth for a child by summer's end. Here green is of the greatest worth, the desert land's rarest color: Islam's standard in some areas.

Our earth, orbiting in the immense aridity of space, has been rightly compared to a miraculous oasis, the nearest thing to paradise a nomad of the cosmos might find in his wanderings. Like the Persian garden, our planet at its best can be viewed as a blissful hideaway and a place seemingly created for joy and fellowship, a model for the hereafter.

Wherever we travel, from tall tropical forest to the Sahara, nature provides us with occasional hints of glory—of promised lands with vineyards and bowers. But such heady visions from humanity's childhood and adolescence may have had their day. Springtime itself may face

Willowy figures of Persian women, possibly drawn under the influence of Chinese or Japanese art, bend as gracefully as the stalks of bearded irises, carnations, mallows, and other flowers in a seventeenth-century tapestry from the collection of the Cooper-Hewitt, The Smithsonian Institution's National Museum of Design. Threads of silk and metal gleam, evoking springtime. To some observers, this scene suggests the traditional holiday outing on seizdeh, *the thirteenth day of the New Year. The Persian New Year comes in spring, near the time of Passover and Easter.*

extinction. We see many places in this world where hope is dying fast: I speak mainly of the developing world, and especially of the tropics. Each new dawn finds some areas moving, almost perceptibly, away from that condition of life we are from time to time tempted to call Paradise on Earth.

Ever since Eden, men and women have earned their daily bread by the sweat of the brow. Wealth or fame is never assured, but honest work well done brings a splendid spiritual reward in itself—a welcome beatitude. Yet merely to survive, hundreds of millions of rural people in the developing nations are unwittingly contributing to the destruction of the natural world around them—source of the very resources necessary to free them from poverty and hunger. Within the lifetime of our children and grandchildren, many jungles, rich wetlands, and fruit-covered plains may come to resemble battle zones. In the Sahel and elsewhere the deserts are already reaching out to claim more territory.

In some ways humanity may have succeeded too well: nature has finally been conquered and we are busy with mopping-up operations. Grenville Lucas surmises that we have tipped the scales against the green world at some point within this century. While the uphill struggle "against" nature was difficult, if we are effectively over the hill, our slide to the bottom could be easy, sudden, and disastrous.

Sometimes humanity can recognize a dangerous situation for what it is. We are also blessed with a little time to salvage part of our biosphere by helping nature's miraculous and forgiving powers of renewal to gain greater importance in our lives.

A great part, perhaps the greatest part of that power resides in the genetic diversity of plants. From it spring our botanical drugs, foods, structural materials, fibers, fuels, and many other classes of goods. Perhaps they've all become commonplace through familiarity, but even everyday magic is not to be despised. If we've learned anything from mankind's experi-

A stained glass window in Canterbury Cathedral portrays the banished Adam delving in the soil just outside of Eden. He employs an iron spade. This prototypical image of all gardeners is conceived as an allegory, but the Old Testament expulsion from Eden, in fact, could be interpreted as an ancient commentary on the revolution in human society that began with the development of agriculture in the Near East.

Adam accepts the "fruit of the tree of knowledge of good and evil" from Eve—an act that will precipitate their losing what has been termed "sheer bliss." This particularly luminous depiction, "Adam and Eve in Eden," was painted by Lucas Cranach the Elder in 1526.

14

ence, it is that we need all the magic, grace, and miracles that can come our way.

In a sense most biological scientists are in the magic business. All of our different scholarly disciplines deal with nature's wonders, some small and others great and awesome. Scientists tend to view the world in terms of their specialty. The botanists are far from alone in their concern over our planetary fate. Nowadays a growing number of research workers in many fields find that their little corner of creation is being violated by pollution, development, economic exploitation or other man-made influences. While none of us like to think about the unthinkable, it is probably fair to say that within the next half century the tropics could sustain damage equivalent in some ways to a nuclear exchange. Little by little the damage is accumulating. It is happening now, day by day, every day.

Ours is a changing creation—the stars, the earth, ourselves. Through the hand of man our present rate of change can no longer be construed as natural. The green world has the faculty of healing, laying up riches, creating stability. The authors of this book believe that we need to make a fresh alliance with the forces of nature. In an appropriate turnaround, conservation is presented as an ally of sustainable economic production and "no longer viewed as mutually exclusive or at opposite ends of the spectrum" in the words of New Zealand's recent Conservation Strategy.

No unfortunate human being on this earth should be blamed for damage to the environment when he or she struggles for survival of family or self. But we might well examine our own consciences if the situation does not improve. This volume persuades us that we can alter our fate for the better. And though we've exploited and over-harvested with apparent impunity for so long that it almost seems the human thing to do, the magnitude of waste today can only be called inhuman. I have very few statistics and studies to back up my suspicion, but I suggest that many of the world's national difficulties may be ecological disasters in disguise. In the Western Hemisphere, for example, two of the most crowded countries are Haiti and El Salvador. They have also sustained high levels of environmental stress.

As I reviewed the early manuscript pages of this volume, something tugged away at the back of my mind—a hint of *déjà vu*. Then I put my finger on it. In 1864, a distinguished scholar-diplomat from Vermont and a great supporter of the young Smithsonian Institution, George Perkins Marsh, wrote his remarkable book *Man and Nature*. Various authorities credit Marsh with having first awakened our conservation conscience. Leafing through those old pages, I rediscovered his belief that what people harm they can also heal. The threats are still with us, yet far more serious now, well over a century after *Man and Nature*.

These pages echo Marsh's essential optimism. We must respond, lest our precious earthly glimpse of paradise comes to naught. The message of *Man and Nature* and of its lineal descendant, *Our Green And Living World*, confronts us at this very moment. It is both a plea and an admonition: we must change or perish. The hour grows late.

Above, a young Colombian woman holds alpine flowers native to her Andean homeland. The pineapple, right, a New World plant which has spread around the globe, became a symbol of hospitality in colonial America.

J. Miller. Sculp.

The choice of the title World Wildlife Fund was made deliberately so as to include plants as well as animals. But at the time WWF was founded in 1961 it was difficult enough to persuade people that so many wild animals were in danger without trying to make them aware of the growing danger to plants. Cuddly animals had an immediate emotional appeal, flowers and trees might be beautiful but it was too soon to get people to see that many of their species might also die out.

Things have changed dramatically in the intervening years. It has gradually dawned on the public mind that even cuddly animals cannot survive in the wild if they have nowhere to live. From there it is but a short step to appreciate that the most important part of habitat is the tree and plant population. The animals cannot survive without them. This discovery led in turn to the appreciation that plants were very important to the environment as a whole. They convert atmospheric carbon dioxide into plant material and into the oxygen that we have to breathe to stay alive. It hardly needs an expert to notice that modern industry and power stations are producing ever greater quantities of carbon dioxide just at the very moment that the earth's tree and plant cover is being eroded faster than at any time in history. The surplus of carbon dioxide is now forming a layer of gas in the atmosphere which tends to act like the glass in a greenhouse, it lets the sun in but it prevents the heat escaping. Plant destruction, unlike animal destruction, affects weather and climate, and if it continues unchecked it may eventually make human life impossible.

Plants have yet another useful property, many of them are edible and while we may fondly believe that we can depend indefinitely on a very limited number of domestic species, the fact is that nature does not stand still and without the refreshment of the genes from wild species the common food plants might well decline in health and productivity.

Once launched into an investigation of the importance of plant life, all sorts of factors start to emerge. Before the discovery and exploitation of oil and coal on a massive scale, all previous civilisations had to depend on renewable sources of energy. In most cases this was wood and it also happened to be the principle raw material for building, furniture and many other essential items. Looking around the world today it becomes only too evident what happens when renewable resources are exploited faster than they can regenerate. The fact is that many of the deserts of this world are man-made and are on the sites of previous civilisations. The erosion that took place after the destruction of the forests effectively prevented regeneration even after the civilisation collapsed and the human population disappeared.

The chilling factor is that in this age we are exhausting both the renewable and the non-renewable resources at the same time. The prospect for the future generations which will have to cope with such a denuded planet are extremely bleak. If the trees go, the source of energy goes.

Energy, food and building materials are not the only things to disappear with the forests. Medicine will also suffer a very serious blow. There is no reason for ordinary people to know quite how dependent the pharmaceutical industry has become on plants for its raw materials. And that is long before the full medicinal potentiality of the plant kingdom has even been discovered. If the plants go, the medicines go.

I hope that by the time you have read this book you will have begun to appreciate the scope of what can only be described as a crisis and recognize the urgent necessity to develop practical and effective means of conserving those core areas of the world which support the biggest concentration of plant species.

Noah is said to have built an ark to rescue the animals from the flood. What the writer of Genesis fails to record is that Noah must also have collected the seeds of all the plants, because without them neither his animals nor his family could possibly have survived.

1984.

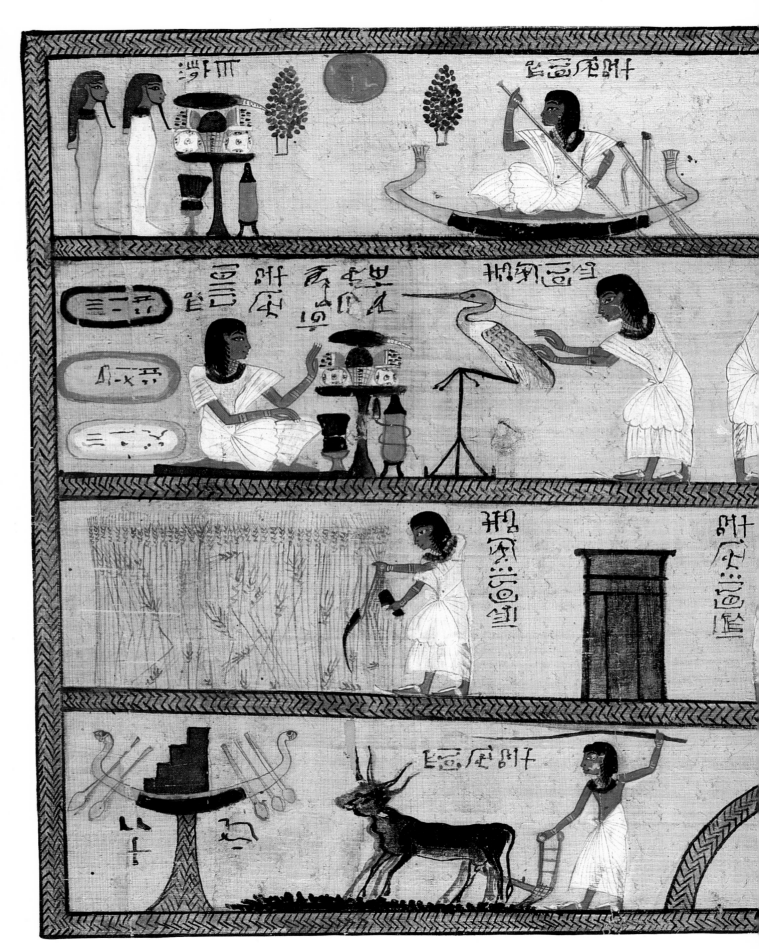

PART I

Tending Our Gardens

ONLY ABOUT FIFTEEN SPECIES—THE STAPLE crops—stand between mankind and starvation. Seven of these are Old World grasses: barley, millet, oats, rice, rye, sorghum, and wheat. Of world cereals, only corn (*Zea mays*) originated in the Western Hemisphere. Wheat and rice make up three-quarters of the world's cereal-grain production, providing half of all human calorie and protein requirements. These grasses, together with various kinds of legume—the beans, peas, and lentils—and root crops such as cassava, sweet potato, and yams, and also with bananas, coconut, and sugar cane, form the bulk of the world's food crops.

Another thousand species or so—the fruit-producing trees, shrubs, and herbs, the vegetables, the potherbs and spices, and those that produce stimulants and intoxicants, beverages, and medicines—add primarily to the quality of life. However, we concentrate for world trade on fewer than two hundred species: quite a narrow economic base when one considers that our planet holds as many as three hundred thousand plant species.

Then there are the plants which provide pasture or feed for the secondary food producers—man's cattle, goats, sheep, pigs, and others. These pasture species and fodder plants are mainly grasses and legumes. To this we must add a few hundred timber species, ones that go to satisfy the ever-increasing demand for wood and lumber for building, for paper pulp production, plywood, veneers, and furniture. Nor must we forget the vast quantity of species whose wood is used for fuel across the world, representing half of all harvested timber.

Primitive cultures did use a much wider range of wild plants. The many were replaced by a narrower range of domesticated species. This decrease stemmed largely from an historical change. We drifted away from a decentralized and low-energy and labor-intensive type of agriculture based on the community and developed an industrial and post-industrial society in which agriculture became characterized by mass

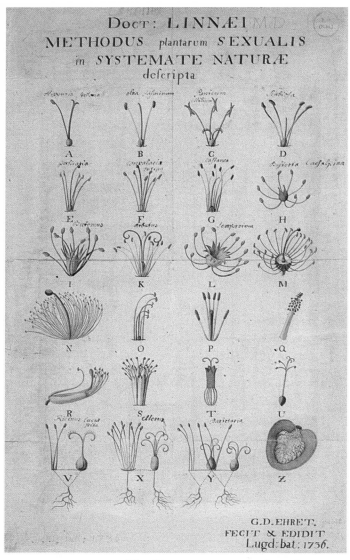

The Linnaean system, based on sexual traits of flowers, vastly simplified the classification of plants brought to Europe from every land.

timber, fuel, and paper pulp, plus a restricted number of fruits, vegetables, spices, and beverages. And if this were all that plants amounted to, we could rationally argue that we now grow all the kinds of plants we will ever need. And provided we maintain a range of genetic diversity of each variety for breeding new races in the future, we could afford to forget the rest. This is very much the kind of argument used by many agriculturalists and economists concerned only with limiting our expenditure on resource management. But it most assuredly would not guarantee our own survival or that of the planet upon which we live, at least not at a level of dignity that would be generally considered acceptable. The reasons are twofold. On the one hand, we are inextricably tied into what some scientists call the "web of life." And on the other, plants represent not only food, sustenance, and shelter, but pervade every aspect of human consiousness.

To give perspective to humanity's intimate involvement with the green world, and how it has changed over the ages—and not always for the better—this first section of the book gives a broad survey of the scientific and aesthetic aspects of plants. The first chapter touches upon the significance of plants in literature, the graphic arts, religion, culture, and society. The second is a pictorial review of plant evolution, with a fascinating glimpse of the co-evolution of the flowering plants and their pollinators. Then we look at the changing fashions and attitudes to plants in our homes and gardens.

production and monoculture.

The survival of the world's expanding populations is made possible not so much by attempting to introduce new crops—indeed no major ones have been added to the list this century—but by gradual and continual improvement of existing crops through selection and breeding. The process has lead to the introduction of new cultivated races (cultivars) well suited to modern mechanized agribusiness.

We could take a narrow view of plants, as many do, considering them merely green machines that somehow provide our essential food. In such a case, the Green Kingdom is reduced to an array of staple crops for food, of fodder plants for our animals, a selection of trees for

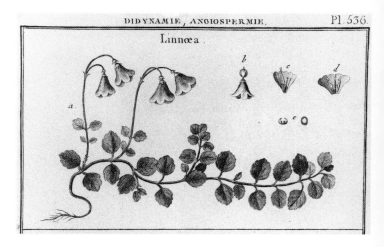

From his Flora Lapponica, *of 1737, Carl Linnaeus holds a Lapp drum. His favorite plant,* Linnaea borealis, *appears right and above.*

M. Hofman sel.

The Benign Connection

Fanciful art captures the notion of a pictured people—either with paint or tatoos—who ruled in Scotland during Roman times. Called Picts by the Romans, these fierce folk left behind elaborate symbols cut into stone. Their monuments are scattered across Scotland.

ACCORDING TO ISLAMIC TRADITION, THE Prophet Mohammed provided an intriguing insight into our aesthetic relationship with plants. "If you have two loaves of bread, sell one and buy narcissus," he advised. "In this way you nourish both the body and the soul."

Such delight in plants—almost an instinct—seems altogether natural when one remembers that, like almost all other animal species, ours evolved as a user of plants. Yet of all species, ours alone learned to plant seeds and bulbs to help assure a plentiful supply of food for immediate needs and for the future. Flowers came to symbolize the cycles of life and renewal upon the land. Archeological evidence suggests that Neanderthals may have placed blossoms into the graves of their departed, a custom echoed by our sending flowers to the sick and wreaths to funerals.

From the earliest days of agriculture—the selection and growing of certain plants to help ensure a regular supply of food—the Stone Age people changed their way of life and consequently began to change themselves, settling down to become farmers and cattle breeders. As civilizations developed, so did the number of ways in which plants came to touch our lives and our sensibilities. As in antiquity, altars in many parts of the world bear flowers or food. To this day, the material and the spiritual are often joined in the taking of food, especially in fellowship with others, ranging from intimate restaurant meals to state banquets.

Farmers, gardeners, botanists, conservationists, and many others find delight in this simple yet profound fact of life: our physical survival depends upon the food we eat and its reliable supply. Yet we often forget this fundamental truth until news of famine or hunger, especially if suffered by millions of children in the less-developed lands, reaches through to us.

In that moment we realize something is wrong, something basic has got out of balance. A growing number of people are beginning to realize that a closer alliance between ourselves and the productive capacity of the green world

Floral tributes to the new-born Christ appear in the famous Portinari altarpiece painted by Hugo van der Goes in 1475. The artist included bearded irises, columbines, violets, and dianthus from the Uffizi, in Florence, Italy.

24

This fresco from a tomb in Thebes, now in the collection of the British Museum, depicts geese, ducks, and the colorful sunfish of the Nile in a garden pond. Papyrus and other plants grow at the water's edge while fruited date palms, orchard trees, and perhaps cedars flourish in what is possibly a walled compound of the Eighteenth Dynasty, 1500 B.C.

is essential if the whole of the world's population is ever to attain enduring prosperity. Let us not forget that half the world faces starvation every day.

As the American publisher Robert Rodale expressed it, "I am convinced that there is a benign connection between the plant world and the human world which we have yet to appreciate to its fullest extent."

In our consideration of what we might call the aesthetics of plants, let us first turn our attention to ritualistic and religious observance. In the celebration of "harvest thanksgiving," found throughout the farming communities of the world, people give thanks to their gods because survival has been assured for another year. Hymns, in nearly as many languages as there are in the world, have been sung in praise of the harvest season. Poetry—only a step away from the hymn—is inspired by the cycle of nature. Just how easily the words create the image of plants in their seasons which many people seldom experience except through the mind's eye is conveyed by John Keats in his "Ode to Autumn":

Season of mists and mellow fruitfulness!
Close bosom-friend of the maturing sun;
Conspiring with him how to load and bless
With fruit the vines that round the
*   thatch-eaves run;*
To bend with apples the moss'd
*   cottage-trees,*
And fill all fruit with ripeness to the core;
To swell the gourd, and plump the
*   hazel shells*
With a sweet kernel; to set budding more
And still more, later flowers for the bees,
Until they think warm days will
*   never cease;*
For Summer has o'erbrimm'd their
*   clammy cells.*

Dylan Thomas provides modern poetry with a mystical yet almost scientific portrayal of a natural cycle:

The force that through the green fuse drives
*   the flower*
Drives my green age; that blasts the roots
*   of trees*
Is my destroyer.

The Bible and other religious works abound with references to plants in verdant lands, with perhaps none so familiar to Christians as the "green pastures" of the Twenty-Third Psalm.

Over thirty-five hundred years ago, the female Pharaoh Hatshepsut sent a botanical expedition down the Red Sea to Punt, the land of incense, and perhaps into the Indian Ocean. This stone relief from her mortuary complex portrays the triumphant return of the voyagers, several of whom bear papyrus and other plants and plant products.

Below, an artisan repairs a wood block used to hand print wallpaper at the English firm of Sanderson and Sons. The printing process requires as many separate blocks as there are applied colors in the finished pattern. Each block is separately inked and printed in sequence. To ensure that the designer's intent is faithfully realized, the individual woodcuts are precisely aligned before the impression is made. Another method, machine printing with a roller, produces such patterns as "Blackthorne," right, designed in 1892 by William Morris, whose work led to the Art Nouveau movement. The plant is also known as sloe, and flavors an alcoholic beverage. Although his renowned firm has long been out of existence, the Morris patterns—especially the flowered wallpapers—have enjoyed continuing popularity. In his day, Morris would not have permitted his design to be reproduced by machine.

From the written word we move to plants in art. A measured brush stroke from China or Japan brings us leaves and blossoms traced thinly yet boldly across silk. Although so many of us have become town-or city-dwellers and divorced from the countryside, we surround ourselves with reminders of nature. Plant and animal motifs abound in our wallpaper patterns, curtain materials, prints, furnishing fabrics and carpet textiles.

Islamic artists gained inspiration from plants for the ornamental calligraphy and exquisite stylized patterns found on manuscripts, book inlays, carvings, pottery, and tiles. The stylized flowers and foliage of rugs and other textiles of the Middle East are often arranged as in gardens. Persians, in their Shiite sect, continue ancient traditions of the direct representation of plant and animal forms, often intertwined in lovely patterns that evoke a sense of perfection in nature and the green habitat.

In Europe, Hugo van der Goes's amazing Portinari altarpiece, now in the Uffizi Gallery, Florence, reveals plants as they may have been picked from the fields and gardens of the fifteenth century. Undoubtedly it represents the artist's belief that Paradise itself will contain such beauty—a beauty far transcending that beheld by the eye. Eden somehow asserts itself in each perfect blossom.

One of the most pervasive of man's links with nature, dating back thousands of years, is the garden itself. As Elizabeth Moynihan writes:

Almost universal in human experience, the concept of Paradise in which man transcends his frail human condition, has persisted while many of the civilizations which adhered to it have disappeared. Belief in it has lessened the pain of life and the fear of death.

The paradise theme appears in the Sumerian cuneiform tablets from the fourth millennium B.C., the earliest examples of writing, antedating even the old Old Persian from which we gain our word paradise. For Sumer, the seedbed of world civilization, gave us the word Eden—one of only two Sumerian words to enter directly into English. The other word, abyss, evokes the darkness of soul which might arise in the absence of the life-giving heat and light of the sun, without the plants which sustain, cheer, and even cure us.

The tradition of the heavenly pleasure gar-

Enamelwork on gold from nineteenth-century Persia conveys the Iranian people's fascination with flowering plants. Iris and tulips are native to this part of the Middle East, and are especially prized because of the brevity of their blooming season. Roses also grow in abundance. Though many Moslem artisans refrain from creating close likenesses of people, animals, and plants, the Shiite Moslems of Iran maintain an ancient tradition of realistic artistic portrayal.

den extended from Mesopotamia through Asia to India of the Moghul era. Its influence slowly spread throughout the world—to Spanish mission gardens of California and to the landscaped gardens of Su-chou in China, Venice of the East.

Evidence suggests that such pleasure gardens were preceded by collections of economic or medicinal plants—herbs and spices, fruits and vegetables. Indeed it is likely that most decorative plants in the gardens of antiquity were also prized for medicinal or culinary qualities.

In addition to local plants, exotic ones from distant lands were grown. In about 1500 B.C., Egypt's Queen Hatshepsut sent her collectors to the land of Punt, probably either Somalia or Yemen. They returned with herbs, trees, and other materials for growing in the gardens of Thebes. Piles of myrrh, a fragrant tree gum of much favor in antiquity, were depicted in stone reliefs. Naturally mummified roots of plants—perhaps from the great expedition itself—have survived in the powdery soil of flower beds of Hatshepsut's funereal edifice at Deir-El-Bahri near the Valley of the Kings.

That the early Egyptians and Babylonians were celebrated for their knowledge of medicines can be ascertained from papyrus inscriptions and clay tablets which have survived. There is also evidence of a herbalist tradition in China dating from around 500 B.C. and continuing largely uninterrupted to the present. In India, knowledge of herbal plants and drugs is recorded from about the same time and was probably based on earlier oral accounts. The Greek herbal tradition of Theophrastus and of Dioscorides was incorporated in the *De Materia Medica* which appeared about A.D. 60. This work became the standard medical reference of the Mediterranean world for more than twelve hundred years.

Herbs and medicinal plants grown in hospital gardens of the Arabian empire after the eleventh century foreshadowed the creation of the physic gardens of western Europe in the sixteenth century. Pharmacy itself, as a profession separate from medicine, was an Islamic institution which began in the ninth century. Luca Ghini, considered the earliest Western pharmacist, founded Christendom's first botanic garden at Pisa seven hundred years later.

To this day, many Muslim herbalists, pharmacists, and doctors seek remedies directly from wild plants rather than from chemical laboratories. Although the use of fresh or dried plants

A luminous emblem of the Italian Renaissance, Botticelli's "La Primavera" transcends ordinary experience, whispering with classical allusion. Though the figures appear static at first, the sinuosity of line suggests movement in both the people and allegorical forces of nature.

Botticelli's masterwork has been described as having captured the restless spirit of an age. The springtime that it evokes may indeed be Europe's resurgence after the Dark Ages. Newly cleaned, today "La Primavera" sheds its light at the Uffizi Gallery in Florence.

Humanity's virtually universal inclination to enrich even the most prosaic urban interiors with green and flowering plants is poignantly illustrated in "The Lost Genius," by the late nineteenth-century American artist Henry Alexander. Geraniums like the one so carefully tended and supported in the old cobbler's shop populate windowsills from Alexander's native San Francisco to Scandinavia. Even—perhaps especially—people who now live and labor in the austere modernity of skyscrapers often introduce a bit of greenery.

Frescoes from Sigiriya, in Sri Lanka, convey a sense of serenity amid tropical abundance. Both of the women hold water lilies, a Buddhist symbol of the soul—a blossom-like glory rising above reflective waters but drawing sustenance from the mud. Exports of tea, rubber, and coconut provide Sri Lankans with one of Asia's highest standards of living.

has diminished in most parts of the West with the rise of the pharmaceutical industry, nearly half of the prescriptions sold in the industrial nations contain chemical agents derived from plants. Modern science affirms the efficacy of the active medical ingredients found in plants, and many synthetic drugs reflect the chemical blueprints of natural substances, particularly the alkaloids and steroids. When the natural product is employed, it is often because the substance extracted from plant tissue is less expensive than that derived from chemical retorts.

A renaissance of interest in herbal remedies and folk medicines has emerged as part of a wide movement toward organic foods and a "return to nature." Several countries retain herbal traditions and, at the same time, seek the best that modern scientific medicine has to offer. Most of the world's people, when they need health care, must rely on herbal practitioners.

Gardens, too, have evolved along different lines over the centuries—with some designed for pleasure and and relaxation, others as centers for research and plant introduction. A remarkable example of the latter is provided by the great network of botanic gardens and stations established by the British across the world. The Royal Botanic Gardens at Kew was a clearing-house for information and material.

During the past 150 years the small domes-tic garden has been a very important part of many people's lives. The plants themselves poured in from many parts of the world, attracting intense interest. Amateur horticulturalists became a new force to be reckoned with and in many countries inspired generations of biologists. This interest has reached even greater heights today, with gardens still bringing delight, as do home greenhouses, window boxes in apartment blocks and even humble pots of house plants.

It is a curious paradox that this vast development of amateur gardening, which involves millions of ordinary people in many parts of the world, has not been accompanied by any real appreciation of the many ways in which plants and their habitats in the wild are being destroyed or modified. One of our aims in this book is to demonstrate this vital link, as survival itself may well depend on how well we cultivate our gardens: great and small, wild and tame.

A Thai flower seller floats her wares—an affectionate tribute—to Princess Maha Chakri Sirindhorn, eldest daughter of King Bhumibol Adulyadej and Queen Sirikit. It is the occasion of the two hundredth anniversary of Thailand's monarchy, celebrated in 1983.

From the Beginning

AMONG THE RAREST OF FOSSILS are flowers in either stone or amber. Yet the bits and pieces that have come down to us sometimes suggest entire landscapes, even Eden's. The most ancient rose we possess, represented by the fossil leaves, right, lived at a time appropriately called the Eocene, or "Dawn Epoch." Smithsonian scientists obtained this particularly fine compression from a deposit at Florissant, Colorado, a world-famous source of detailed impressions of ancient flowers, leaves, insects, and other delicate specimens.

At least chronologically, nature might seem to have exchanged dinosaurs for flowers. Hardly 70 million years ago, now extinct dinosaurs and plants dominated the earth. In the blink of an eye, geologically speaking, new plants spread across the land. They were the angiosperms with their abundant flowers. They provided the welcoming bower where the mammals, insect forms, and the human race itself evolved.

However engaging may be the new menagerie, and such ultramodern plants as the orchids,

from the earliest days of life on our planet, a fundamental chemistry has applied. Chlorophyll, the catalytic pigment of most green plants, maintained its constant production of foods and fiber. The fungi did their work of recycling these complex organic materials into the simpler chemicals which green plants require for their growth. The algae, especially those of the oceans, consumed carbon dioxide and gave off oxygen, the breath of life for animals.

Only a few decades ago, algae, fungi, bacteria and green plants

Insects, birds, and even small mammals are attracted to particular plants and thus serve as pollinators and as agents of dispersal for seed, as with the finch atop a thistle, opposite. Egyptian artwork at right, may represent a fig wasp, an insect whose life cycle is intimately involved with that of the fig. The pollinating behavior is intricate. To produce the large Smyrna fig of commerce, which has no male flowers, branches with wild caprifigs (with male flowers) are tied to the trees. Female fig wasps transfer pollen from the Caprifigs to the inflorescences which will ripen into the edible fruit. Inside the green fig, the female wasp places eggs within tiny fig flowers. The fruit thus ripens at the same time the wasp eggs mature and hatch.

were considered all part of the Plant Kingdom, now called the Plant World. Instead of three categories of matter—animal, vegetable, and mineral—we now have six, including higher green plants; fungi; algae; and blue-green algae and the bacteria. The latter are the most primitive forms in the Plant World.

Examination of the evolutionary tree-of-life reveals some intriguing facts. The Animal Kingdom is sketched in, sometimes with broken lines, as stemming from the Plant World. If such schematics were taken literally, we might say that in primordial times the ances-tors of algae and of human beings were one and the same. That era is so remote, however, that such speculation is not liable to generate the kind of controversy stirred up by Charles Darwin's assertions of man's common ancestry with some prehistoric simians.

One small but important aspect of the common ancestry of today's plants and animals does, however, apply here. Many plants and many animals do share certain metabolic processes, often called pathways. Some yeasts, for instance, produce insulin and other hormonal substances that closely resemble those secreted by humans and other animals. The human brain also releases chemicals called endorphins. Beta-endorphin closely resembles alkaloid substances found in opium, and both the human and the poppy versions become tightly bound to the same chemical-receptor sites in the brain—as if some chemical key were inserted into a biological lock. Recent scientific speculation points to a plant counterpart of interferon, a substance in the blood which may suppress the action of viruses and retard the development of some tumors. In any event, human interferon appears to cure viral infections in plants.

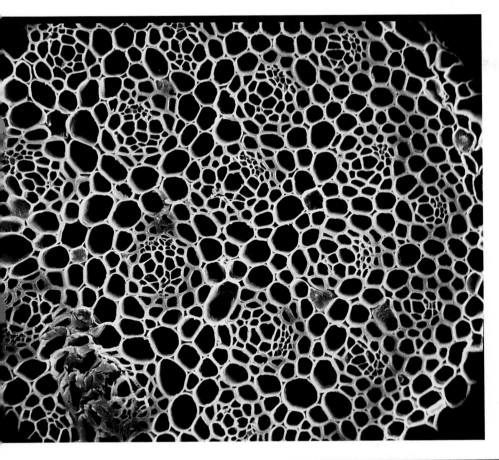

Fossil walnut leaf and nuts were extracted from an Eocene deposit at Clarno, Oregon. The original material of the nuts has been replaced by a form of quartz known as chalcedony. Such trees and fruits belong to the great class of flowering plants, called angiosperms, and they arose not long after the Age of Dinosaurs. Conifers are much older, and more primitive, with ovules exposed on cone for chance pollination by wind. Pollen, below, is blown about easily because of its size and bladder-like wings. Whether ancient or modern, higher plants possess a vascular structure, shown at left in cross section. One set of tubes transports water and dissolved minerals from the soil to the leaf. The leaf produces sugars which reach all parts of the plant through a second set of tubes.

There is also the case of the jojoba plant which produces an "oil" almost identical to that from sperm whales, a unique parallelism in chemistry.

Scientists have hardly begun to explore this exotic corner of organic chemistry, and coincidence rather than scientific correlation may be at work in these few cases. Yet in such small ways, and in larger ones, human needs are exceedingly well served by substances fashioned through the alchemy of the Plant World.

There has long been a folk belief that each plant possesses a special use, if we can but find it. This belief has been one of the "proofs" for a faith in a Divine Providence. While discussion of religious beliefs is far beyond the scope of this book, from a scientific point of view there do seem to be subtle and creative relationships not only

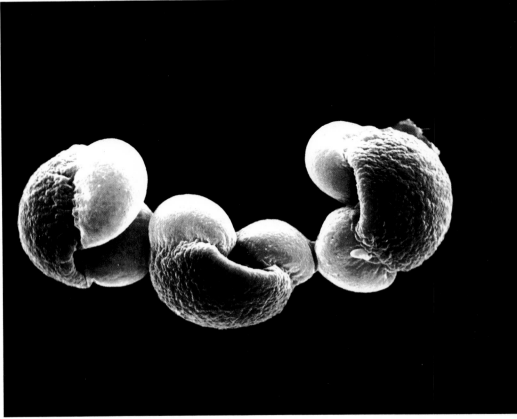

Like much flora and fauna of large isolated land masses, the duck-billed platypus of Australia and the lemur of Madagascar have adapted to specific habitats which are now gradually dwindling. In the early 19th century drawing by Ferdinand Bauer, above, the platypus (Ornithorhynchus anatinus) pair gambols beside one of the water ferns common in their semi-aquatic domain. The apparently dazed stare of the wide-eyed lemur belies the tenacity with which it grasps its insect prey.

All of the approximately 60 Marsilea *species have leaves with four lobes, giving them a resemblance to four-leaf clover. In shallow water like that of the platypus habitat the leaves stand upright above the water's surface; in deeper water, they float.* Marsilea *species occur in temperate and tropical areas of the world, the one below being native to the island of New Caledonia in the southwestern Pacific. In contrast to terrestrial ferns, which produce spores on the underside of their leaves,* Marsilea *species bear their spores on separate stalks.*

between plants and animals, but among the six natural kingdoms—mineral, animal, plant, fungal, algal, and bacterial including the blue-green algae.

In particular, the plant kingdom proper and the fungal kingdom represent two important parts of the nutrient cycle; one contributing nutrients to the benefit of the other, round and round *ad infinitum*. In the chapter about tropical rain forests, beginning on page 98, a discussion can be found on the particular importance of root-mat fungi called mycorrhiza to the survival of the jungle's great trees.

Considering the almost unbeliev-able diversity of species of the flora and fauna, what strange and awesome claims of the jungle herbalist may next prove to be true? What cures for cancer or for afflictions of the nervous system will be discovered? We cannot predict but we already know, in a sense, that almost miraculous cures and other benefits flow from the juices and sap of many plants.

In the past, for instance, we have depended for medicine upon plants which have the right proportions of healing substances. Now we know that many dozens of alkaloids can occur in a single plant, and that a very scarce one can be the healing one—the case with the chemical vincristine, an exciting cure for childhood leukemia, see page 186. With sophisticated analytical equipment, scientists are just beginning to unlock the chemical storehouses within plants, and to tap a source of biological wealth which has been building since life began.

Tropical katydid from Sabok, Malaysia, stands on its head. The meaning of this gymnastic pose is uncertain but may be a warning to predators, a sexual invitation to a mate, or even mimicry. Thus arrayed, the insect resembles an orchid, just as the orchid, opposite at top, resembles an insect.

In the Venezuelan jungle, the orchid Pleurothallis raymondii *is adapted to resemble the female of its pollinating insect. This is one of the numerous highly evolved syndromes that allow plants to reproduce. It is called pseudocopulation by scientists. The male insect mounts the flower, attempts to copulate, and in doing so spreads pollen from another orchid to fertilize this one.*

At left appears one of the Plant World's most remarkable examples of co-evolution. The Lepanthes calodictyon *orchid of Colombia displays a sexual structure that closely resembles a small fly. In effect, the plant is displaying a tiny artificial lure to attract a male pollinating insect. So intricately wrought is this sexual bait that entomologists attempted to classify it until they discovered that it was plant, not animal. Right; grasses like these have no need for elaborate blossoms. The wind carries their pollen just as it may transport some of the tiny spiders that have formed a colony between grass stalks.*

A very ancient order of insects, the beetles have evolved many spectacular markings to rival the glorious blossoms which they frequent. When true flowers evolved, beetles were the earliest pollinators. The beetles ate the pollen, rather than sipping nectar as do modern insects that visit flowering plants. Scientists suspect that beetles simply plowed through the sexual parts of ancient plants and promiscuously spread the pollen as they moved. As we have seen, pollination is less chancy today, with insect and flower often fitted together through co-evolution. About forty percent of all insects are beetles, including the snout beetle Chloropholus bioculatus from Madagascar, above, and the diamond beetle from Samar in the Philippines. Opposite; no beetle at all but a true bug of the order Hemiptera, it seeks the golden center of a flowering Vellozia of Brazil. As with beetles and other kinds of insects, this knobby creature's coloration tends to match that of the flower on which it spends much of its time.

Old and New

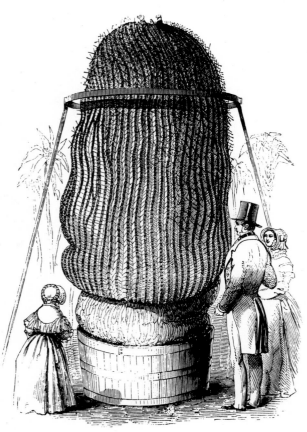

"MONSTER CACTUS," AT THE ROYAL BOTANIC GARDENS, KEW.

*Opposite, proper Victorians intrigued by exotic flora stroll through leafy arches in the Palm House of the Royal Botanic Gardens, Kew, ca. 1850. One of many botanical curiosities on display in the Gardens was the giant Mexican barrel cactus (*Echinocactus platyacanthus*) or "visnaga," above.*

JUST PAST THE TURNSTILE AT THE ROYAL Botanic Gardens, Kew, the visitor enters a floral microcosm. Like other major botanic gardens, Kew has endeavored over the centuries to build up a representative collection of the world's flora, capturing a small measure of "earthly paradise."

Many of the exotic plants introduced into Kew during the last century required the protection afforded by the conservatories and palm-houses constructed in Victorian times. In the heyday of these great glass temples, many of which were to be found even in the private gardens of the rich patrons of horticulture, a wide range of tropical plants could be appreciated. People were able to gain a sense of wonder at the immense diversity of the botanical world, especially of the tropics which so few of them could ever visit.

Although there are far more and cheaper facilities for travel to the tropics today, we must still turn to the few large botanical gardens if we wish to see the tropical plants that our great-grandparents favored. These include *Sabal* palms, plants of enormous proportions with huge fruit-stalks and gigantic fan-shaped blades. Most people are unfamiliar with the common denizens of those temples of tropical flora. They were the heliconias, the giant white bird-of-paradise (*Strelitzia nicolai*), the useful pandanus (*Pandanus utilis*), and stately traveler's tree (*Ravenala madagascariensis*), an opulent plant rivaling the ornamental palms.

Changes in taste, economics, and domestic style also have much to do with what we grow and why. Interest in many of the parlor plants grown in Victorian times has waned. Some formerly popular varieties of geranium, China rose, heliotrope, and jasmine have been allowed to die out for lack of interest, although it is ironic that attempts are now being made to rescue such plants from oblivion. New varieties have largely replaced the old standbys, but an ever-increasing range of exotic house plants is grown today. The cut-flower vogue has changed and greatly expanded, so much so that for countries such as the United States, many plants must be grown abroad.

Colombia accounts for more than 90 percent of these cut-flower imports. This Andean land near the equator is famous for its roses, standard and pom-pom chrysanthemums, and its enormous production of carnations, over four hundred million of which are flown to Miami

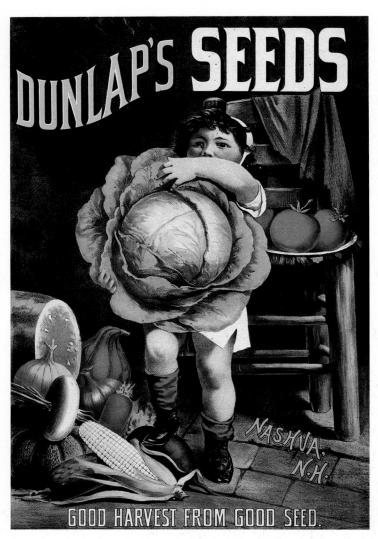

*Attractive packaging, above, helped
Victorian and Edwardian seedsmen
create an enormous public demand
for their wide variety of high-quality
vegetable and flower seeds.*

famed for its exports of bulbs such as daffodils, tulips, and freesias as well as for its cut flowers and house plants. Northern and Central Europe are in turn often supplied with flowers such as roses and carnations from the Mediterranean countries and the Canary Islands.

Seed companies tend to import from countries where the climate favors good formation of seed and labor costs are low, the latter factor being especially important in the production of seed from those hybrids which require pollination by hand. Perhaps 90 percent of the hybrid petunia seeds sold in North America come from Central America. Guatemala, Japan, Kenya, Mexico, Chile, and Taiwan supplement California's large output of seed destined for the American market. Many of Great Britain's celebrated seed firms have their crops grown under contract in Mediterranean and other warm countries where maturity can be assured. Seeds are an ideal item of commerce, often very valuable in relation to their weight. The seed trade offers splendid opportunities for developing countries to exploit new markets.

The early 1980s witnessed an increasing demand for wild flowers in the United States where shade tolerant plants such as impatiens and begonias were also popular. The reason: trees and bushes planted during the domestic construction booms of the fifties and sixties began to mature, thereby adding shade to the properties of millions of Americans.

Foliage plants, garden seeds, and cut flowers constitute only a part of the plant business. The trade is ever more lucrative due to the demand by discerning collectors for rare exotics. Many cacti and orchids, as well as palms, succulents, carnivorous plants, and alpines are propagated by commercial growers. Since so many of these varieties have been decimated in the wild by private and commercial traders, the Convention on International Trade in Endangered Species of Wild Fauna and Flora (CITES), was established in 1973. By 1984 there were 85 signatories. This agreement allows commerce in listed species of the world's flora to be regulated by means of permits issued by the country of origin and, in some circumstances, the country of destination.

Most exotics reach the United States through Miami. Of the more than one hundred and sixty million living plants of all kinds imported each year, 10 percent are listed in CITES. Half a million of these are orchids, distributed amongst a thousand species, arriving principally

each year. In 1982, a recession year, people in the United States spent three billion dollars on cut flowers, a large proportion coming from South America.

Though Florida supplies 80 percent of the tropical foliage plants sold in the United States, most of the country's domestically grown seeds and cut flowers originate in California. During the last decade, international offerings have made inroads on the California market, and dependence on foreign sources is increasing due to higher fuel costs for maintaining greenhouses.

Of the half-billion tea roses purchased by Americans in 1981, a tenth were imported. On a world scale, Israel specializes in miniature carnations and roses, while the Netherlands is

from suppliers in Mexico, Belize, Honduras, India, and Thailand.

Ninety-eight percent of the seven to ten million cactus plants imported each year come from Mexico, Canada, the Dominican Republic, and Japan. Overall, six hundred species arrive from more than 50 countries. Most exported specimens go to Japan, West Germany, and the United Kingdom. Many Latin American cacti are first brought into the United States and then exported to Europe and elsewhere. In 1976, for example, about half of the fifteen thousand cacti exported from the United States to the United Kingdom originated in Mexico and South America. Ninety percent of the cacti exported from Belgium to Britain in 1977 were originally imported from the Americas, and 70 percent of cacti exported from West Germany to Britain were imported in the first instance from the Americas.

Both legal and illegal cactus collecting takes place in Arizona, Texas, California, and New Mexico. Arizona cacti may be legally collected from private land once a permit from the owner has been obtained. Large specimens are excavated using hay forks, chains, hoists, and hydraulic lifts from the back of trucks, and sold to waiting customers who often specify in advance their preferred plant size.

In plant transportation, as in all matters, necessity stimulates ingenuity. Below, the sealed glass Wardian case, precursor to the terrarium, permitted the safe long distance transport of delicate and valuable plant materials. Bottom; special equipment moves an enormous Mexican Agave atrovirens, *with padded leaf-tips and towel-swathed flower stalk, on the grounds of Kew Gardens.*

THE GIGANTIC WATER-LILY (VICTORIA REGIA), IN FLOWER AT CHATSWORTH.

Above, seven-year-old Annie Paxton bravely proved the strength of the giant leaf of the Victoria amazonica *water lily in the Duke of Devonshire's conservatory at Chatsworth in 1849. Later, Sir Joseph Paxton designed the engineering landmark Crystal Palace in Hyde Park, using the ribbed leaf of the water lily as a structural model.*

In spite of a stringent Arizona state law which prohibits the collection of desert plants without a permit, up to $1 million worth of cacti may have been stolen from the land in 1978 alone. Efforts to stop thefts of saguaro, prickly pear, cholla, and barrel cacti have earned threats for some officials. Even the murder of witnesses has occurred. Once in a while the green world strikes back, however. After heavy rains in the Tucson area in late 1983, tree-size saguaro cacti absorbed so much water that their roots could barely hold the plants upright. A saguaro toppled over and crushed a cactus rustler to death.

The soil, climate, and availability of labor in California make fields such as the one near Lompoc, shown above, a major domestic source for the United States' production of flower seeds and cut flowers. Right, a turn-of-the-century seedsman's advertisement suggests the range of flower species available to home gardeners of that era.

DUNLAP'S

PHLOX
DRUMMONDS (FINE MIXED COLORS)
BLOSSOM THE FIRST SEASON
PRICE 10 CENTS
A.H. DUNLAP & SONS

SWEET PEAS
FINE MIXED COLORS
BLOSSOM THE FIRST SEASON
PRICE 5 CENTS
A.H. DUNLAP & SONS

VERBENA
FINE MIXED COLORS
BLOSSOM THE FIRST SEASON
PRICE 10 CENTS
A.H. DUNLAP & SONS

CONVOLVULUS
NEW CRIMSON VIOLET
BLOSSOM THE FIRST SEASON
PRICE 5 CENTS
A.H. DUNLAP & SONS

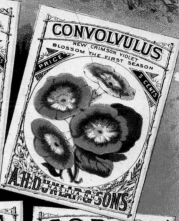

CARNATION
FINE MIXED COLORS
BLOSSOM THE SECOND SEASON

CANDYTUFT
WHITE
BLOSSOM THE FIRST SEASON

CACALIA
FINE MIXED COLORS

COBŒA

FLOWER SEEDS

NASHUA, N.H.

PANSY
GERMAN FANCY COLORS
BLOSSOM THE FIRST SEASON
PRICE 10 CENTS
A.H. DUNLAP & SONS

BALSAM
FRENCH DOUBLE MIXED
BLOSSOM THE FIRST SEASON
PRICE 10 CENTS
A.H. DUNLAP & SONS

COCKSCOMB
GLASGOW NEW PRIZE
BLOSSOM THE FIRST SEASON
PRICE 10 CENTS
A.H. DUNLAP & SONS

PRIZE ASTER
RECENT FLOWERED GIANT MIXED COLORS
BLOSSOM THE FIRST SEASON
PRICE 10 CENTS
A.H. DUNLAP & SONS

CALENDULA
NEW METEOR
BLOSSOM THE FIRST SEASON
PRICE 10 CENTS
DUNLAP & SONS

DATURA FINE MIXED
CANTERBURY BELL
COREOPSIS
CHRYSANTHEMUM FINE MIXED COLORS BLOSSOM THE FIRST SEASON
CATCH FLY MIXED COLORS

A SPLENDID ASSORTMENT.

PART II

Global Mosaic

IN 1802, THE GREAT GERMAN SCIENTIST AND plant explorer, Baron Alexander von Humboldt, having travelled from the Orinoco River to the Andes of Ecuador, scaled the peak of Chimborazo then thought to be the highest mountain on earth. He realized that in moving from sea, to river basin, to high desert, to peak he had passed through each of the earth's climatic realms—almost literally from fire to ice. From such observations and insights modern plant geography and ecology were born.

Humboldt's expedition raft, here with the stern full of botanical specimens, rides near Guayaquil. From a vantage point far above that afforded by Chimborazo, the view from earth-orbiting satellites, gives the impression that even the Himalayas appear to be flattened out. In effect, we behold a mosaic of biological zones, or biomes, with ocean and desert often appearing to dominate the picture. On a few fortunate days, with clouds temporarily dispersed, we can even behold the rich green of Amazonia.

Two-thirds of the total surface area of the globe is blue sea. A third of the remaining surface area, the land, is either arid or semi-arid. Polar and montane ice buries an additional 11 percent of the land. Tundra—or "barren ground," as it is unjustly termed—covers another 10 percent. Little more than 40 percent of the land area supports grasslands, forests, wetlands, and their transitional zones.

Thus we can see how incredibly slight is our green margin—our margin of survival—how incredibly rich in its millions of species. Earth is both womb and cradle. However sparse, vegetation is a moist coverlet that protects us; which we, in turn, would do well to cherish. The endlessly rocking seas nurture their own special flora and fauna.

Differences between life zones can be striking. Yet in a sense the land biomes form a continuous series. Hyper-arid deserts shade into relatively moist areas, some holding forests of giant cacti. Grasslands emerge, some supporting trees. The seasonal but scarce rainfall of many

Mediterranean lands is sufficient for tough-leaved evergreens, a stage in the progression toward the evergreen and deciduous woodlands of the well-watered temperate areas. Finally we encounter the biological exuberance of the high tropical rain forests. Swamps, marshes, and bogs have their own particular (and rather restricted) faunas and floras, the latter including mangroves, marsh grasses, and water weeds.

Plants of the vast oceans—usually algae, some gigantic and others microscopic—proliferate only where nutrients are present in particular concentrations and where temperatures and light levels allow. In overall productivity, the coldest waters are often the richest. Coral reefs of the warm oceans provide pocket habitats of

great biological activity. Scientists discover islands to be miniature worlds, some with the same range of biological zones as a continent or the whole planet, from sweltering seashore to glacier-capped peaks.

Wet and dry, cold and hot, high and low—in unique combinations, the range of natural circumstances dictates the evolutionary adaptation of plants and animals and the ability of the land or sea to carry its cargo of species. In each part of our planet's living mosaic, humanity alone holds the power to alter the historic processes of habitat destruction which could consign our species and so many others to the fossiliferous stratum forming at our feet.

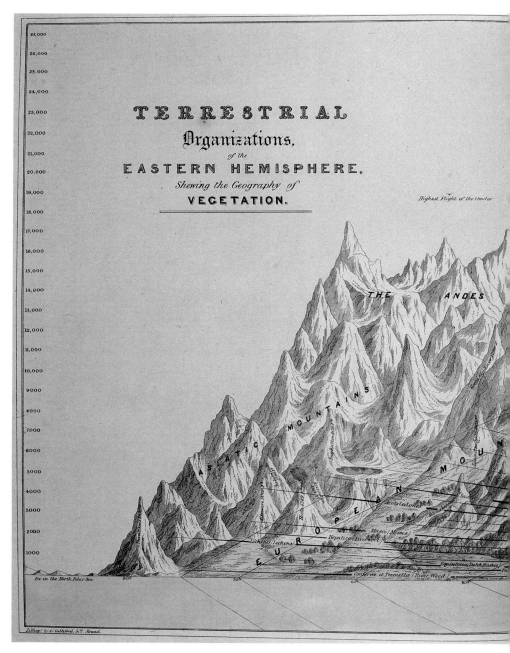

Prevailing winds from 12 points of the compass buffet a rotating globe in an old German book illustration. It depicts nature's seasonal forces, symbolized by the progression from youth to age. The movements of global patterns of air pressure over the earth's surface produce such climatic phenomena extremes as doldrums, monsoons, and fierce polar winds, many of which are related to the shifting of the earth's axis between the summer and winter solstices and the attendant latitudinal migration of the wind belts.

Baron von Humboldt promoted the concept that plants form communities according to latitude and altitude. The notion that widely separated areas can have almost identical growing conditions and thus quite similar, though not necessarily related, plants and animals inspired many other naturalists. T.B. Loader, an Englishman, compiled a schematic of botanical zones, at left, for the Eastern Hemisphere. Such pioneering efforts gave rise to the modern science of ecology, which today includes such specializations as "plant sociology," the scholarly study of the species composition of plant communities. Note the speck below Everest, which represents the Andean condor's highest flight.

Deserts

deserted. Neither are they always hot. Arctic and Antarctic weather can be as dry as the Sahara's. So tenacious is life in the northern polar regions that splashes of colorful lichens and carpets of miniature flowers occur even on the northernmost speck of land, tiny Kaffeklubben Island off Greenland. And in Antarctica, algae live literally within rocks whose crystalline structure retains moisture and allows light to penetrate.

Furthermore, the tropics are not always wet. Even in the moist tropics, there are some areas where mountains stand in the way of prevailing winds and thus create rain shadows. Also, some porous tropical soils dry out rapidly and some areas are seasonally arid. Essentially, though,

Though remote, Afghanistan has suffered much conquest. Genghis Khan and Alexander the Great passed this way, where archeological workers delve in history's dust, once irrigated and productive.

there are two parallel and globe-circling desert zones. The northerly one includes the Sahara, Sahel, Middle East, southern Asia, and the southwestern United States. The southerly one covers southern and eastern Africa, Australia, and dry portions of Peru, Chile, Argentina, and Brazil.

Whether bizarre or beautiful, desert plants brilliantly illustrate the extraordinary adaptability of living things. In deserts, plants and other living things must live strictly within the means imposed upon them by their surroundings. To survive, some plants must complete their reproductive cycles during the few days or even hours of sufficient moisture. Even size is an adaptation; some desert plants are so tiny that the botanist must lie flat on the ground to study them. The so-called resurrection plants such as Rose of

Jericho (*Anastatica hierochuntica*) reflect desert conditions. They go dormant and appear parched and dead until the rains revive them and they start growing again.

Like the cacti of the New World, the many succulent euphorbias of Africa's Sahel and Ethiopia have evolved from species with "conventional" leaves to ones with thick skin and spongy stem tissues that store water. In both euphorbias and cacti the true leaves are often reduced or rudimentary.

Once the desert bloomed in Afghanistan's Sistan region, below. Fine towers and walls rose at Shar-i-Gholghola, but crumbled as irrigation canals fell into ruin.

This is testimony to the influence of climate, and of gradual changes in climate, upon the outward appearance and inner processes of plants. For plant material as diverse as lily, amaryllid, cactus, and euphorbia, adaptations to the arid environment often result in the oddly-shaped, tree-like species widely sought by collectors. As already noted, some are currently protected under provisions of the international CITES treaty, which prohibits traffic in threatened species without a permit.

Strange succulents have also evolved in southern Africa, the Kalahari, and Namib Deserts. Their popularity as house and specimen plants has also led to efforts to protect them in the wild. The Namib Desert of Angola, South West Africa (Namibia), and South Africa receives less than two inches of rain per year and like amounts at the coast as dew from sea fogs. Yet the bizarre gymnosperm *Welwitschia bainesii* has succeeded in this unlikely setting. It has a long taproot, cones, and only two leathery, strap-like leaves which grow continually throughout the life of the plant, a span measured in centuries. From an evolutionary standpoint, this region seems to specialize in cucurbits, aloes, and euphorbias.

The island of Madagascar holds both moist tropical jungle and an arid zone with plant families found nowhere else on earth. Some species resemble a cross between a pipe organ cactus and a yucca. Perhaps the strangest, *Alluaudia procera*, looks like a bent and thorny telegraph pole. It grows to 50 feet in height, and clusters of flowers pop out of the column. Along with other unique species, it is threatened by the introduction of Mohair goats, which can eat the young leaves with their sharp thorns!

Like the plants, people of dry lands must also live within the desert's means, but we need more to eat than thorns. Unlike plants we cannot survive on dewdrops, sunbeams, and soil minerals. Yet even in deserts, when the population is low, the scant vegetation sustains wandering tribes and, rarely, settled people.

The greatest danger may come from too much, rather than too little, rain. Encouraged by variable rainy spells, rats, rabbits, locusts, and cattle multiply quickly. Grazing animals crowd the edges of newly-filled water holes and devour the thin green haze of plant life.

Under this onslaught the sparse vegetation cannot slow water or protect the soil. Falling rain causes a variety of erosion problems. In its rapid run-off, the rain is seldom evenly absorbed by the thirsty soil. Erosion destroys the habitat that plants need. As dryness returns the increased herds eat everything that sprouts. They eat the flowers before the plant can go to seed. Old seeds in the sand—nature's saving's bank—sprout and are immediately consumed. All biota eventually disappear.

While some plants and animals can hibernate, people cannot. In some cases flight is the only alternative to death. The populations of many sub-Saharan cities are already swollen by the influx of ecological refugees, a category of displaced person found in nearly every life zone both wet and dry and particularly in the tropics.

Proper management of plant and water resources gives us the best chance to avoid loss of arable soil. Desertification is a word coined to describe the encroachment by sand, and the deterioration of soil in zones where ecological change deprives the land of its ability to sustain agriculture and human existence. In Africa the

Roller coaster turns of the Trans-Andean Highway cross a dry divide to link Santiago, Chile, and Mendoza, Argentina. Once virtually impassable, many desert regions can be crossed now in several hours. A vestige of the past, right, a caravan of sturdy mountain cows treks a primitive road in the Vākhān, a strip of difficult territory which formerly linked Afghanistan to China.

Sahara does not often "invade" the villages. More typically, aridity begins at the village perimeter and moves outward. Understanding the terminology and mechanics of this process is less important than recognition of the extent and seriousness of the problem. In fact, most people only need remember the word Sahel.

Africa does have dust bowls. During the past century Africa's deserts have grown by more than three hundred and eighty thousand square miles, an area equal to the size of Egypt. Arable land nearly half the area of London is lost to the desert each year. A transition zone around the southern Sahara, the Sahel encompasses parts of the Sudan, Chad, Niger, Mali, Algeria, and Mauritania. The dry spell that brought such tragedy to the Sahel in the 1970s lingered and then intensified in the early 1980s. The long term harvest of wood in the forest belt south of the Sahel has created ecologically unstable conditions in parts of Ghana, Ivory Coast, and other countries near the Bight of Benin.

Overleaf; Andean desert lies at the foot of Mt. Chimborazo. Fruiting Opuntia *is in the foreground, with tree cactus at left and* Agave *at right, with pointed leaves.*

By studying such areas, we learn that irregularity of rainfall is the key condition governing the sparsity of vegetation and its vulnerability to human influence in arid zones and transitional areas. As rainfall becomes a bit more periodic, seasonal, or reliable, deserts grade into semi-arid lands. Such a transition occurs between Cairo, Egypt, and Tel Aviv, Israel.

If it were not for the Nile, Egypt would be as deserted as the Saharan highlands to the east and west of the great river valley. Indeed, visitors to Egypt take delight in being able to stand with one foot in desert and the other in lush field or meadow—the limit of irrigation. At the Ma'adi Sporting Club golf course near Cairo, a

59

Left; pale blossoms belong to a member of the Orobanchaceae, a family of plant parasites without green coloration. Roots of this plant join to the roots of an unseen host, probably a tough desert shrub of the sort eaten by a nomadic Tunisian's camel, below.

perfect English landscape has been recreated right next to wasteland.

Hyper-arid lands are those having a mean annual rainfall of less than one inch; arid lands have up to 8 inches; and semi-arid lands have as much as twenty inches. In many places, if water can be brought in, people can make the desert bloom—though perhaps not forever if ancient Mesopotamia, the Sistan of Afghanistan and Iran, and the Sabean Empire of the Queen of Sheba are any guide.

Sheba was indeed a woman of substance—more than a casual visitor to Solomon's court. If an ancient tale is as true as the environmental message it contains, some of her wisdom was gained from Solomon himself.

Above; botanist Thomas Baines inspects a female Welwitschia bainesii *plant, of Africa's Namibian Desert. Below, an American cactus (left) and an unrelated African* Euphorbia *(right) exhibit parallel evolution of body form—spines for leaves and spongy stems that retain water—though flower form and sap composition are quite different.*

The story, told in Ethiopia's ancient "Book of the Kings," the *Kebra Negast*, recounts a meeting between Solomon and Sheba. The young queen, so bright and headstrong, loved to pose hard questions. Upon her arrival from her Arabian kingdom, she immediately asked Solomon what in all the world was most valuable. He delayed his answer and ordered a banquet, including sweet wines and salty meats, for that evening.

Awakening in the night, Sheba found her mouth and throat parched and her tongue clinging to the roof of her mouth. She left her bed and hurried to the night table with its goblet of water. Solomon stood in hiding, and pulled her

Flowering plants brighten the deserts of North America. Below, Abronia villosa, *a member of the four-o'clock family, spreads across Mojave sands. Bottom, flat spines disguise rare* Pediocactus papyracanthus, *the paper-spined cactus, as a grass. Its habitat includes Arizona, New Mexico, and Mexico. Opposite; a desert in the southwestern United States blooms with mounded* Oenothera, *the evening primrose.*

Sun-baked and cracked into irregular patterns, the soil at Capital Reef National Park in Utah surrenders itself to the harsh heat and aridity of the annual dry season.

hand away just as it touched the vessel. When he repeated her own question, Sheba knew the answer—*the most precious thing is moisture in a dry land.*

Sheba went home to construct water conservation projects, perhaps even Yemen's great dam at her capital city of Marib. Upon her legacy, as the story goes, was built the power of Saba, an empire of influence in both Africa and Arabia and the forerunner of Ethiopia.

At the very least, the old story dramatizes the grave responsibilities of those who rule in desert lands today. Barring dramatic climate changes for the whole planet, however, we do possess the means to control desertification of our precious arable land. When dry spells do come, people can be far better prepared. Future droughts need not be so damaging as those of the present and of the recent past.

Egypt's Western Desert raises a rocky shoulder to help create the Valley of the Nile. Plant growth ceases at the limit of irrigation. Architectural detail at Jerash in northern Jordan dates from Roman times. In antiquity the settlement lay along the spice route from Southern Arabia to Damascus. Richly laden with frankincense and myrrh, caravans from the Queen of Sheba's realm passed this way.

Grasslands

FOR MANY OF US, SAVANNA MEANS THE wooded grasslands of East Africa, with its antelope herds, rhinoceroses, elephants, wart hogs, buffalo, jackals, lions, and scavenging vultures. Here the genus *Homo* evolved from the Australopithecinae some one to three million years ago. Great plains contain Kenya's Amboseli Masai Game Reserve and Tsavo National Park; Tanzania's Serengeti National Park; and the Kalahari and Gemsbok National Parks in South Africa.

These famous grasslands are usually characterized by species of *Acacia*, wonderful spreading thorn trees of lion-and-giraffe country. They generally lose their leaves during the dry season and produce flowers upon the bare branch. After a long dry season, new leaves sprout with the rain. These graceful trees reach 60 feet with a spreading crown twice as wide as they are high. Most have thick bark, grow edible pods, and develop an extensive root system. Some are, as the botanist terms them, myrmecophilous—particularly attractive to ants which feed on them and at the same time protect them from other damaging insects.

Local people periodically set fire to encourage growth of new grass. Whether man-made or caused by lightning, these dry-season fires both suppress tree growth and allow bulbous plants to become abundant. Showy flowers from species such as *Herschelia lugens* (Orchidaceae), *Gladiolus aureus*, *Gladiolus bullatus*, *Hyacinthus corymbosus*, and *Moraea loubseri* (Iridaceae) are vanishing through over-collection for the trade and by habitat modification. This fire is also important as a means of replenishing the soil materials that are leached out by seasonal rainfall. The veld, an expansive grassland in the eastern part of the interior of the Cape Province in South Africa, has been widely given over to the cultivation of cereals and potatoes.

The word savanna (deriving from the Spanish *sabana*) originally applied to the extensive, more or less flat grasslands or prairies of northern South America, Brazil's cerrado and campos and the llanos of Colombia, Venezuela, and Argentina. These areas constitute 1.17 million

A black-maned lion basks in the sun on one of a system of African savannas and other dry grasslands. They form a mosaic of magnificent arid and semi-arid habitats within which dwell an encyclopaedic array of fauna—from aardvarks to zebras.

square miles of barely usable land that hold only 2.5 million people. Now the term also refers to tropical areas of Africa and Australia where tall grass and scattered shrubs and trees grow. In total, dry and semi-arid grasslands support a hundred million people—one person out of every forty on our planet.

The three categories of savanna suggested below, while not all-inclusive, provide some insight into the formative factors. First are those grasslands dominated by long dry seasons. Second are those where long-lasting erosion has made soils too shallow for trees. Third are those where human activity has shaped the land, particularly through the clearance of nearby forests in shifting cultivation marked by the use of fire. Far greater areas of tropical Africa would come into cultivation or be used as pasture were it not for the tsetse fly, the carrier of sleeping sickness to man and domestic livestock, which infests nearly a third of tropical Africa's grasslands.

Continental plains support development of huge grassland systems such as the prairies of North America, the pampas of the Argentine, the steppes which stretch from Hungary and Austria to Siberia, and the veld of southern Africa. Eurasia's grassland steppe is the largest, spanning almost five thousand miles from the Danubian Basin to the Altai Mountains of the Soviet Union, to Mongolia and China.

With the spread of agriculture, the western steppes in Austria and Hungary have suffered a depletion of native flora as have Ukrainian grasslands, largely converted to crops since the 1930s. It is hard to realize that these grasslands came into production much later than the prairies of North America, possibly because red winter wheat was introduced into Canada and the United States from the Ukraine and not vice versa. Today wheat and corn fields claim the lion's share of America's heartland. The Soviet breadbasket extends from the Polish border to the upper Caspian Sea and beyond the Volga.

North America's Great Plains are a puzzle in that the amount of rainfall they receive should by all accounts support many more trees. Some

researchers suspect that fire has impeded a natural biological succession—from grasses and herbs, through scattered woodlands, and finally to old-growth, or climax forests. The same prairie wind that once turned thousands of windmills has fanned tens of thousands of prairie fires in past ages. Some experts believe that throughout the millennia of human habitation in the New World, fire has possibly shaped the development of prairie soils, themselves a secondary product rather than the cause of the grassy surfaces.

It has long been recognized that many fires are started by lightning but ethnographic research reveals that Plains Indians were responsible for others. Doubtless the landscape is vulnerable and even receptive to fire, for we see that grasslands occur chiefly where there are at least occasional periods of dry weather during which

the ground cover dries out, and also where the land surface is smooth to rolling.

Grasslands may also extend into broken terrain adjacent to such plains. Fire is a factor even here. Whipped by dry winds sweeping across areas of low relief, recurrent blazes can check the spread of woody vegetation. In a few instances, suppression of fire seems to allow the gradual establishment of trees.

Were it not for the Amerinds, the original settlers of interior North America, parts of the Great Plains might be covered with both grass and trees. Even in the early days of European settlement in North America, only islands of trees appeared midst the sea of grass. Frequently— perhaps annually—the Indians burned the prairie. Apache, Arapaho, Cheyenne, Kansas, Omaha, Oto, Pawnee, Ponca, Shoshoni, Sioux, and Ute: why should these tribes want to set

George Catlin's 1832 painting "Prairie Bluffs Burning" evokes the drama of a windblown prairie fire, with fleeing deer. Such fires may have contributed to the scarcity of trees in the Great Plains of North America.

fires in the first place? Perhaps it was done to improve hunting by rousing or stampeding game animals; or to create pasture; to enhance visibility; to collect insects; to increase the yields of seeds, berries, and wild vegetables; to remove trees in favor of other growth; to clear land for planting; to aid warfare; to facilitate travel; or simply to produce a spectacle.

Motives may seem less important than the likelihood that for fifteen thousand years or more, native Americans employed a tool of great leverage— fire—to radically alter the natural surroundings.

In contrast, other investigators believe that the dearth of trees on the Great Plains is due primarily to geological and geographical factors. The mixed prairies, those located between the tall-grass and the short-grass prairies, date back to Tertiary times, about 25 million years ago. Active uplift of the Rocky Mountains precipitated changes in the prairie climate; in particular, rain and snowfall were reduced.

As the mountains rose, they intercepted moist winds from the Pacific Ocean. Clouds heavy with rain or snow dropped their burdens as they rode up the western face of the mountains. The air had been largely dried out by the time it reached the plains, creating a rain shadow, a dry region beyond a mountain barrier. This is part of the reason why the Middle West of the United States is often subject to low summer rainfall and cold, dry winters.

It is assumed that tree growth was difficult in such a climate, and that the earlier, original, forests gradually disappeared. Grass and other herbaceous plants flourished in their place. Thus, as some researchers see it, our prairie has evolved from an aboriginal one which came into its own after the Mesozoic Era.

The details of the various scenarios may seem less important than a basic fact of more recent North American geography and history: within the past century and a half, people have altered drastically a work of nature which endured with only gradual modification during the

six hundred and fifty thousand centuries since the dinosaurs.

The modern progression from the occupancy of prairies by the Indians before the 1830s, through a century of settlement and development, led to the Dust Bowl of the 1930s. The 1980s bring renewed dangers of erosion from wind and water.

The defeat of the Sauk and Fox tribes in the Black Hawk War of 1832 brought the end of organized resistance to prairie settlement before the American Civil War of 1861-1865. Over-grazing and trampling by the settlers' herds, plowing for farming, along with a rising influx of "sodbusters," caused the greater part of the Middle West's well-drained prairies to be plowed, planted, or converted to other uses.

The area of Illinois, Iowa, and Missouri is now the Corn Belt. Iowa alone produces 20 percent of the United States corn crop and 15 percent of the soybeans. The air traveler can look down to a vast, nearly unbroken sweep of mile-square "sections" imposed upon the prairie, a gridwork of fields projected in the earliest surveys and plans. This grand conception was a fitting prelude to agribusiness.

All the years of agricultural production have brought much good to the United States and to the world. Yet the conversion of prairie to cropland created soil instability. On the heels of an international depression, which had brought intensified use of croplands, came an upheaval of nature in the 1930s, the "Dirty Thirties." High winds from the south whipped the dried soils of Texas and Oklahoma into immense dust storms, making a "Dust Bowl" out of five states. Some of the dust, combined with topsoil and sandy particles, was blown across the Mississippi River, beyond New York City, to fall on ships in the Atlantic Ocean. One estimate indicates damage to more than two hundred and eighty million acres of farmland.

The soil problem fell to Franklin Delano Roosevelt, originator of the shelterbelt in the United States. The scale of the program that created those protective rows of tough tree species, also called windbreaks, was described by *Los Angeles Times* reporter Bryce Nelson:

More than 222 million of 'FDR's trees' were planted in the Prairie States Forestry Project from 1935 to 1942. From the Dakotas, these shelterbelts stretch south 1,000 miles into the Texas Panhandle in a

Winds roaring across the Great Plains blast an exposed farmyard, below. At right, farmstead buildings at McCabe, Montana, huddle within the protection of four rows of windbreak trees. The great plains of Siberia also hold windbreaks. The People's Republic of China has established a belt of tough Australian pine (Casuarina equisetifolia) *along eighteen hundred miles of coastline. Recent reports suggest that protective rows of trees may also soon be planted around the capital city of Beijing.*

200-mile-wide belt in the transitional area between the Tallgrass Eastern Prairies and the Shortgrass Western Plains.

The United States Soil Conservation Service, part of the Department of the Interior, had spread the word. Dust Bowl conditions eventually subsided and the farms became viable during World War II. Many windbreaks have survived into the eighties. They augment other soil conservation methods including terrace planting, strip cropping, and crop rotation.

Yet the Dust Bowl threat is with us still. Up to a third of the three hundred and twenty million acres of farmland in the United States suffer erosion from the force of wind and rain. *Washington Post* staff writer Ward Sinclair has pointed out that major soil modification occurs in the Blackland Prairie of Texas, the Corn Belt of Illinois, Iowa, and Missouri, and the Palouse Prairie of eastern Washington, Oregon, and Idaho. The Palouse region, where wheat has only recently replaced prairie grasses, annually loses from fifty to a hundred tons of topsoil an acre, far greater than the norm of five tons. Some localities in the Plains States have topsoil 20 feet deep, with four to five feet of friable loam common elsewhere. So it is particularly difficult to

interpret the impact of losses except in areas of marginal soil.

Although the breaking of the tough prairie sod in the 1800s is recognized to have led to the Dust Bowl damage of the 1930s, farmers today are plowing marginal soils, breaking former pasture land, and felling windbreak trees to expose more land to cultivation. Farmers speak of the need to expand acreage to pay for new machinery and other expensive "inputs" required in industrial agriculture. More crops must be harvested just to break even when grain prices fall, as has recently happened. Whatever the motivation, neglect of basic conservation measures has

become apparent in many places. Sound approaches to sustained production, which have been proved and improved upon for many decades, are being thoughtlessly abandoned.

In the developing countries, the problems of using wisely the resources offered by tropical grazing lands are even more critical. As pointed out in a recent Unesco/UNEP/FAO report (*Tropical Grazing Land Ecosystems*), the balanced rural development of such grassland zones lies at the heart of their national development strategies and programs.

Mediterranean Shores

WE RECALL LEARNING AS SCHOOLCHILDREN that the benign climate of the Mediterranean—the mild wet winter and hot dry summer—favored development of human settlements and the transition from man the nomadic hunter to man the sedentary farmer. Here on the fabled shores arose great cities, countries, and empires. From the lands around the sea came successive waves of civilization—the Egyptian, Phoenician, Etruscan, Roman, Hellenic, Persian, Arab, Byzantine, Jewish, Christian, Provencal, Spanish, Ottoman and others. These have largely shaped our attitudes and ethics to the present day.

Not only is the Mediterranean Basin identified with many of the great values of Western culture, but it is probably true to say, with Di-Castri, that nowhere else on this planet has man been so closely and intimately associated with the environment. Perhaps nowhere else has the land so deeply influenced man's behavior and culture: in turn these have shaped the past and present landscapes. And of great significance, perhaps more than anywhere else in the world, the modification and degradation suffered by the Mediterranean landscape has been caused by man's misuse of the land rather than by nature's processes.

Part of this pattern of misuse came with the spread of agriculture; and this region was one of the first to develop land for farming. From this center of origin and dispersal came such crops as wheat, barley, lentils, beans and chickpeas, fruit trees such as figs, and the pistachio, olive, cherry, and almond. Agriculture played a major role in modifying the landscape through the clearing of trees and shrubs for field crops, grazing of sheep and goats, terracing of slopes, and irrigation of arable lands.

Perhaps most important of all was the deliberate and accidental use of fire with its devastating effects. Clearing of the land for crops, combined with the felling of forests to provide timber for shipbuilding, led to the characteristic landscapes of today's Mediterranean—cultivated fields, orchards, vineyards, olive groves, grazing lands, and scattered forests of evergreen oaks, pines, cedars or firs. Much of the landscape that we associate with the Mediterranean is therefore

This first-century mosaic from Israel depicts a goat-like animal cropping a tree branch, one factor in the degradation of ancient forests into scrubland.

Sentinels against the sky, a few conical trees and some columns remain. Once slopes of the Mediterranean Basin held dense forests where classical civilizations flourished.

man-made, or at least substantially modified by man's activities, which have also helped to create the extensive scrubland called matorral.

In humid and sub-humid zones of the Mediterranean, some of the high matorral is capable of reaching 12 to 16 feet, notably with *Erica arborea* and *Arbutus unedo* in Corsica or the *Arbutus andrachne* maquis of south Anatolia, thus preventing typical oak forests from re-establishing themselves. The soils are well developed and similar to forest soils in the same regions. Such unusual climax maquis is apparently of considerable age, the diameter of the trunks suggesting four to five hundred years.

High matorral refers to what French authors call "maquis," or the "macchia alta" of Italian, "matorral denso" or "espinal" of Spanish, and "chaparral" in English literature, and especially when applied to California or the Southwest of the United States. This dense formation of scrubby growth is well adapted to dry spells and contains low evergreens—both trees and shrubs—and may be dominated by plants with broad leaves, and with or without pines or junipers. In Chile such underbrush is also known as "matorral," but "mallee" in Australia, and "renosterveld" in the Cape province. Low matorral is also called "garrigue" or similar names in the Romance Languages. Greeks call it "phrygana" and other eastern Mediterraneans say "batha." It is a dwarf scrub of dry slopes and hillsides.

After so many centuries of habitat modification, little remains of the original evergreen oak forests. Writing twenty-four hundred years ago, Plato referred to the mountains of Attica in Greece where only traces survived of the vast forests that once covered the slopes. Writing four hundred years later, Pliny asserted that a squirrel could swing from branch to branch the whole length of the Iberian Peninsula.

From the Black Sea to Alexandria to Gibraltar, in antiquity this was very much a land of

wine, honey, and olive oil. Here nomadic hunters settled down to become farmers. Here, at the "middle of the world," three continents meet and thus the attitudes and ethics of much of mankind were shaped by events at this hub of geographical and historical development.

In climatic and thus in biological terms, we find a transitional biome, a life zone fitting between the dry tropics and the temperate regions. Furthermore, the classical lands and the world's several other lands of mediterranean climate are of recent origin, having appeared for the first time in the Pleistocene Epoch only ten thousand years ago. All are strongly influenced by cold ocean currents. Despite their prominence, these lands are transient, having reached their maximum extent in historic times.

The mediterranean climate has a pronounced winter rainfall alternating with a distinct summer drought at the time of maximum temperatures. Winter usually brings downpours. Rapid runoff limits the amount of moisture reaching the roots of plants. The flora is adapted to withstand drought—the primary factor limiting the development of wild and domesticated plants. Leaves are often thick and leathery, inrolled, densely hairy, or in other ways modified to prevent water loss in unfavorable circumstances and many of the plants are spiny or thorny.

There are, in fact, many ways of defining mediterranean climate and the many subdivisions of it. A useful one, based on the proposals of the French botanist Emberger, divides the year into one or several types according to the length of the summer drought—from as long as eleven or twelve months to as short as one or two months. If the extremes are excluded, we have four types—arid, with nine to ten months of drought, semi-arid with seven to eight months, sub-humid with five to six months, and humid with three to four months.

In addition, there are at least six different rainfall patterns that can be recognized and these have great biological significance for the plant life concerned. The patterns are characterized by four letters which indicate the seasons **A**utumn, **W**inter, s**P**ring, **S**ummer in order of decreasing rainfall. These are: WAPS with the highest rainfall in the winter, followed by the autumn and with the driest season in the summer. This pattern is found in places as far apart as Algiers, Palermo, Istanbul, Lisbon, Beirut, and Perth. WPAS, another in which the highest rainfall occurs in the winter, is widespread in

east Mediterranean countries such as Syria, Lebanon, Australia, Jordan, and in parts of north Tunisia, Italy, and Algeria.

PWAS, with the rainfall highest in the spring, and significant precipitation in winter, is characteristic of parts of Turkey including Ankara, of Morocco with Marrakesh, and parts of the east Pyrenees. PAWS, with the highest rainfall in spring and the next highest in autumn, occurs in parts of Turkey; in Italy including Assisi; France with Carcassonne, and Guercif, Morocco, which typify this particular balance, or regime, of moisture and temperature. APWS is a pattern with the highest rainfall in the autumn followed by the spring and occurs in parts of France such as Marseilles, Aix-en-Provence, Montpellier, Narbonne, and in parts of Morocco and Tunisia. AWPS is the final pattern. It exists in such coastal towns as Grasse, Monte Carlo, and Antibes and in parts of Algeria and Tunisia.

Landscape and History

Grazing, agriculture, fire, lumbering, pollution, and urbanization are the main agents of modification of the Mediterranean environment. Human influence was first strongly felt during the Neolithic Revolution when agriculture appeared in the Near East. Archeological evidence documents the cultivation of wheat and barley, peas, beans, and lentils from 9000 to 7000 B.C. The breeding of cattle began about the same time, although goats and sheep were kept earlier. In the western Mediterranean, agriculture and livestock were common by 3000 B.C. The climate dried out during the period of animal introduction—Europe's Copper and Bronze Ages. Shepherds, burning the forests to create artificial pastures, imposed further pressure for change on the dense, primordial vegetation.

As the distinguished Spanish forester Luis Ceballos wrote, "the destruction of forests is fatally linked with the history of mankind." The clearing of trees and shrubs for field crops, grazing of sheep and goats, terracing of slopes, and irrigation of arable lands all figured in the modification of the Mediterranean landscape. Most important was the effect of deliberate and accidental fires.

Of course some forest fires are of natural origin. Well adapted to fire, many plants could reestablish themselves after the occasional catastrophic and almost explosive conflagrations of the drier parts of the Mediterranean Basin. But this resurgence could rarely be sustained on lands which would thereafter be subjected to the plow or to close grazing.

In times of peace and war, fire—the first force of nature—was often turned against nature itself. Flame was used to extend arable lands, to enhance early pasture growth for grazing. The burning of forests and fields was common as a reprisal or to smoke out the enemy during military campaigns; but even scorched-earth campaigns are often limited in scope. Pressure on woodlands and other vegetation probably eased during epidemics, famine, and wars.

Coniferous forests, especially those of *Pinus halepensis*, are particularly susceptible to flame. And the matorral are rich in essential oils which readily burst into flame. Even today, fires sweep across nearly eight hundred square miles of forest and scrubland each year. Not only does this represent a major loss of wood, but the direct and indirect costs are serious. Fire fighting, fire prevention, regeneration of forest, damage caused by erosion, silting, and general damage to soils and vegetation must all be taken into account. It is impossible to fully assess the overall effect of fire on the forest and scrub, or matorral vegetation, but the damage can be estimated at hundreds of millions of dollars a year.

Grazing became a great economic force in antiquity and, with agriculture, helped to create an enduring economic system. Now we find literally hundreds of millions of sheep, goats, donkeys, horses, cattle and camels in the Mediterranean Basin. With the exception of productive forests, any natural vegetation can serve as pasture. Worse yet, the nature of the Mediterranean's climate can actually encourage overgrazing. Unlike temperate regions where snow and frost protect the vegetation for a season, here animals are able to forage during both winter and summer. In hot weather, plant growth nearly ceases and any cropping can be fatal to the survival of plants.

Nonetheless, in the Mediterranean a family's sheep and goats are usually regarded as a very important investment and the grazing pattern is firmly entrenched. Goats are particularly well adapted to survive on the Mediterranean's leathery vegetation. First, they contribute to the destruction of the forest and then thrive in the degraded scrub.

On a more positive note, Mediterranean lands still possess an abundance of native forage species. The commercial exploitation of legumes, grasses, and other plants of enormous potential

Detail of both the female cones and the smaller male reproductive structures of the Cedar of Lebanon (Cedrus libani) are shown in this drawing by George Dionysius Ehret, a master botanical artist of the early 1700s. The art is structurally incorrect, in life the two sexual features appear on different branches. Biblical and other sources indicate that Solomon obtained cedar beams for his temple in Jerusalem from King Hiram of Tyre, in Phoenicia, now Lebanon. The Egyptians incorporated cedar into their mortuary temples and used cedar oil in embalming. Centuries of commerce in the fragrant and strong cedar wood have decimated the once luxuriant forests of Lebanon's mountains, and only about four hundred of the old trees have survived in their native habitat.

*A single layer of the bark of these Portuguese cork oak trees (*Quercus suber*) will be carefully stripped. Numbers painted on the trees guide the cutters. A new layer will have formed in about a decade and the process can be repeated without damage to the tree. The cork bark, which is part of the tree's adaptation to dry conditions, has a combination of useful characteristics unmatched by man-made materials.*

could help to bring destructive processes more into balance with productive ones. Scientific investigation of underexploited species followed by economic development and introduction, generally come slowly, however.

Mediterranean vegetation remains rich in trees, if we are speaking of the number of species which survive. Comparing forest areas of the past with those of today produces a very different evaluation, for it is difficult to find more than occasional stands of well-developed forest.

About a hundred species of trees remain. Forest of cork-oak and various other evergreen oaks probably reached their greatest extent in prehistoric times as they spread into semi-arid zones of the region on various soils. Matorral, often with scattered oak trees, comprises a third of Mediterranean ground cover today.

Various oak forests occupy lands between the drier parts of the Mediterranean zone and the wetter temperate zones. They are especially important in Turkey, where they represent more than a third of all forested lands. Four deciduous species predominate. The Kermes oak (*Quercus coccifera*) rarely forms forests except in parts of the east Mediterranean, where it is regarded as a separate subspecies. Mostly it grows in the scrubland of the western and eastern Mediterranean. Cork-oak forests survive in North Africa, Portugal, and Spain. The tough,

thick, and highly insulating bark is a classic example of evolutionary response to ecological conditions. No synthetic material has been able to duplicate the remarkable characteristics of this plant product which has been stripped from the trees for centuries to "cork" bottles. Recently we have learned how to press and fuse cork particles and dust into a variety of useful objects.

Pine forests occupy vast areas of the Mediterranean Basin. Except for the montane forests of black pine (*Pinus nigra*), they do not usually constitute the climax, or old-growth forest. Rather, they have gained a foothold in areas of former forests of broad-leaved evergreen trees. In almost all cases the original cover has been destroyed or modified by economic activity. The Aleppo pine (*Pinus halepensis*) and also *P. brutia* form extensive forests in Spain, France, Italy, Greece, Turkey, Tunisia, and Algeria. Both are particularly susceptible to flame, however, and forests of Aleppo pine suffer great damage in the semi-arid and humid zones.

Here, again, we witness an oscillation between destructive and creative forces. Fire, in fact, does play a positive role in dispersal of some pine seeds, and in helping to assure their germination. The cones burst open during forest fires and project the seeds several yards, allowing some of them to escape damage.

The umbrella or stone pine (*Pinus pinea*) and the Maritime pine (*P. pinaster*) grow in coastal regions and near the sea, the latter mainly in Spain. Black pine forests occur in the Atlas mountains of North Africa. Cedars cling to these same slopes and to those of the Taurus mountains of Anatolia as well as in Lebanon and on the island of Cyprus. Winters in these highlands are characterized by minimum temperatures, persistent snow cover, and frequent fogs, but the summers are long and dry.

Fir species are scarce, with two surviving species in the east and two in the west. In the late 1960s, a relict forest of *Abies nebrodensis* was found in Sicily. All of the 21 mature trees showed damage from fire and grazing. These have been fenced off, and the Sicilian Forestry Service has established a nursery for seedlings. More than a hundred young trees have been planted near the older survivors. The two eastern species grow near the timberline on mountains in Greece and Turkey.

The mild climate of the Mediterranean has sustained many ornamental plants, trees, and crops from abroad. These include pomegranates and bananas, maize and rice, the prickly pear cactus, agave, aloe, persimmon and date. Palms of many kinds line streets and grow in parks and gardens. Other ornamental trees include the jacaranda, Cape honeysuckle, Indian bean, foxglove tree, tree-of-heaven, pepper tree, coral tree, pagoda tree, cassia, and flame tree. Trumpet vine and bougainvillaea clothe walls with their brilliant colors. These and many more give a tropical luxuriance to the urban landscapes.

Trees from Australia—including various eucalypts, also called gums—have been planted on a large scale in the countryside. Some of these have become naturalized and form part of the "wild" landscape. The acacias, especially *Acacia dealbata*—the mimosa of florists—appear conspicuously on the hillsides of the Côte d'Azur.

Stable stands of native vegetation of the coastline—already broken up by extensive cultivation of oranges, lemons, and other fruit and vegetable crops—have been further reduced through the dramatic expansion of tourism. Concrete hotels rise and apartment blocks sprawl. Development is almost out of control in some localities. Along hundreds of miles of coastline the natural vegetation has been sacrificed to the construction and tourist industries.

The sunshine, salubrious climate, and pockets of floral magnificence of the Mediterranean region create a sense of careless ease and the good life for many visitors. More careful scrutiny reveals an underpinning of poverty, with little hope in some areas for other than transient prosperity around the centers of commerce and tourism. Heavy pollution of air and water place further stress on the environment's ability to cleanse itself. Polluted air around Rome, Athens, and other famed cities literally corrodes the classic statues and monuments which have in some cases endured for thousands of years.

The diverse civilizations of the Mediterranean Basin have become molded, despite the frequent conflicts between some of them, into what DiCastri has called a common cultural unit with a common denominator of shared values, which is based on the co-existence of a great range of different languages and cultures. In other words, the Mediterranean has become a way of life not only for its native inhabitants but for waves of immigrants and tourists. History, culture, tradition, climate, agriculture, and vegetation combine to provide substantial fulfillment of human aspirations. This earthly paradise is fragile, however, and the illusion could be destroyed very easily unless human pressures on the region are carefully monitored.

Beyond Classical Lands

The first Europeans to explore the Cape region of South Africa, central Chile, California, and southern Australia must have been pleasantly surprised, for they found that in climate, natural vegetation, and general appearance of the landscape, these areas resembled parts of France, Spain, and Italy. What is more, settlers from the Mediterranean Basin brought olives, citrus fruits, grapevines and many other trees and crop plants with them. These often grew well, with beneficial consequences.

In due course, the term mediterranean came to be applied in both a generic and a scientific way to five widely separated areas of the world, including California, Chile, Africa's southern cape, and parts of Australia. Similar vegetation and climate are found in a few other areas such as the Caspian Sea's southern coast and uplands near Shiraz, also in Iran. Today, scrubby evergreen trees and shrubs and other tough plants are so widespread as to be thought typically mediterranean. Yet, as in the Mediterranean Basin itself, such shrublands often follow the degradation of various forests. The presence of this landscape in the world's other mediterra-

The seven-hundred mile sweep of California, from Mexico to Oregon, is captured in this mosaic of 29 Landsat images. It has been color-enhanced to highlight the state's various life zones. The pale wedge covering the southeastern portion of the state is the Mojave Desert. The High Sierras, dominated by coniferous evergreens, appear as a dark green ridge running down the eastern central area. Scattered green areas in the north and along the coast are also evergreen woodlands, some of which include redwoods and sequoias. The long pale shape between the Sierras and the coast is the Great Valley of California, an area with a mediterranean climate that irrigation has turned today into one of the most productive agricultural regions on earth. The russet zone surrounding the Great Valley and extending along the coast to Mexico holds grasslands and the famous chaparral, comprising shrubby plant communities of varying species. Chaparral is California's most important ground cover.

This remarkable image was developed jointly by the NASA-Ames Research Center and the California Department of Forestry in 1979 as part of a legislatively mandated assessment of forest resources.

nean areas is very recent indeed, usually dating after European settlement.

California

California possesses one of the most celebrated floras of the world. Rich and diverse, its plants range from giant redwoods, firs, and pines, to the hundreds of colorful annuals that emblazon the state in springtime. Those parts of California and Oregon (and the nearby Mexican State of Baja California) that have a mediterranean climate produce more than four thousand species of flowers and ferns, nearly half of which are endemic—that is, restricted to the area. These are composed of both very ancient species, called relicts, and of ones more recently evolved.

The mountain realm, or montane biome, is represented by the Sierra Nevada where much of the state's forest is found. The cool and moist coastal north, with its redwoods and sequoias, becomes a temperate rain forest in Oregon and Washington State. Thanks to the work of the Forestry Department of the State of California, we can see all the major woodlands in that state in a remarkable mosaic of satellite images which accompanies this text.

Extensive grasslands appear throughout the coastal valleys and foothill regions of California. They are artifacts of human enterprise. The introduction of exotic plants to California began in May 1769 when Father Junípero Serra reached San Diego Bay and founded the first permanent European settlement. Annual grasses and weeds have invaded the original grasslands, with the total number of introduced species calculated at 674, of which 559 are of Old World origin. Three-quarters came from Eurasia and North Africa.

The Indians had long maintained a stable relationship between themselves and the natural world around them. As with other such associations built up slowly over centuries, a balance between creative and destructive aspects had been struck. For instance, the Indians of California judiciously employed fire in much the same fashion as Neolithic hunters and foragers of the Old World.

European settlers dramatically increased the frequency and scale of fires. They cleared forest and chaparral to open up the range for grazing and agriculture. They also cut timber for mine props and housing. Extensive grazing, combined with a primitive rural economy at the time of the Gold Rush of 1849, led to frenzied

Growers in Fallbrook, California, make use of rolling hills to cultivate avocados. Careful terracing helps to conserve the soil and hold water for the small trees. American in origin, the avocado has a high percentage of oil in its buttery flesh.

exploitation of the land. Unaware of their effect on the landscape, the immigrants neither knew nor cared about conservation. They have been described as people out to make the most money in the shortest possible time.

The effect on vegetation was to break one pattern of land use and to establish another. The original grasslands, dominated by perennials, were severely altered. The total area of grassland was vastly enlarged through conversion of forest and shrubland. As a result, the extensive grasslands of California are an artifact created by man in the past two hundred years. The gold is gone, some of the original forests are gone, and many of the original grasses have been replaced by introduced annual species.

History moves on, and the economic emphasis of Gold Rush days has changed. By the end of the nineteenth century, with increased standards of living, the old system of dry farming lost importance. The grasslands on the slopes were too difficult to manage with machinery. Abandoned fields often reverted to scrubland.

Today, mostly in the south, shrubland covers more than thirteen thousand square miles of foothills and lower mountain slopes. This is the famous chaparral, made up of tough evergreen bushes which form impenetrable thickets and are so well adapted to the long dry summers. The term itself is derived from *chaparro*, Spanish for a scrubland of evergreen oaks. Now it indicates any of various kinds of dense scrub or brushland. The dominant one is called hard chaparral and occupies the dry zone above the coastal sage scrub, or soft chaparral. The latter supports Californian sagebrush, yellow pine, and several other conifers.

Though chaparral looks much the same throughout its extensive range, the composition varies. Several hundred species are involved, most of them woody. It takes a specialist to tell the difference and to interpret local conditions that favor one particular mix of species. In each instance, this growth represents a stage in the regeneration of the old-growth, or climax vegetation, which it replaced. Protected from passing economic enthusiasms, the succession from scrub to forest can occur. Appropriate plantings could hasten the return and enrich the land and its inhabitants. Fortunately we know how to repair old damage so often done in haste and out of ignorance.

Of little or no commercial value, chaparral nevertheless forms the state's most important watershed vegetation. In a sense it is priceless. Its management is of vital interest since the damage or removal of chaparral leads to serious erosion and floods with enormous economic consequences. A great deal of research is devoted to the study of its biology and the effects of fire.

Chile

The natural vegetation of Chile has been modified by human activities through many centuries, a process begun long before the arrival in 1541 of the Spanish soldier Pedro de Valdivia, the original European settler. To cultivate various crops, the native Chilean Indians cleared the land with fire. There are even traces of shifting agriculture in Neolithic times. The European settlers introduced livestock, which put great pressure on the land. There was also extensive harvest of wood for charcoal, especially from the lower slopes of the mountains. Trees were also cut and burned to provide additional acreage for crops and grazing. All this enterprise led to severe depletion of dry forests and shrubland.

Chile's mediterranean vegetation occurs in the populous central heathland, the Valle Central, where Valdivia settled. Similar conditions apply in the "Norte Chico" or Little North. Brushland called matorral occurs on the slopes of the Coastal ranges and foothills of the Andean Cordillera, while the Valle Central includes a degraded savanna called *espinal*. A montane mattoral of low evergreen shrubs has emerged on dry slopes above the scrubland, while on some coastal terraces a different type of brushland exists. The mix of species and their relative proportions are sensitive indicators of local environmental conditions.

The honey palm once flourished in the foothills of the coastal ranges of central Chile, but today only a few stands of the majestic *Jubaea chilensis* remain. Mature trees are still felled to extract syrup from the sap. The practice dates from the seventeenth and eighteenth

Native and introduced flora have been mixed in a park in Pretoria, South Africa, right. A painting by Marianne North, opposite, depicts natives of the ancient and distinctive Cape Floral Kingdom, including wild geraniums. The coastal belt of South Africa is also rich in Protea species, shown growing in a large colony below. Cape flora, like that of other regions of the world, is increasingly threatened by burning, grazing, and the introduction of alien flora.

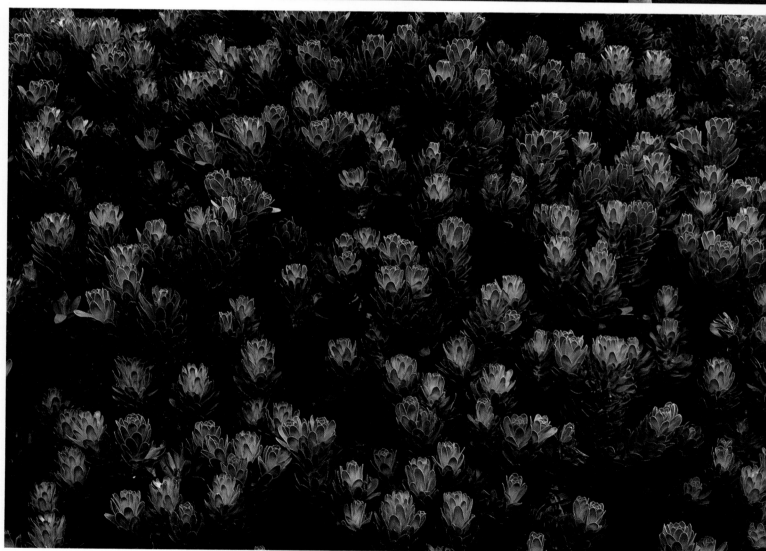

centuries and may accompany the honey palm into extinction in much of its native habitat. Palm honey once provided sugar for the whole of Chile, the methods of production being noted down by Charles Darwin during his travels through Chile in 1834 and 1835. Regeneration of the palm is restricted by grazing. Because of its shiny gray trunks, up to sixty-five feet tall and nearly eight feet in diameter, it has been planted in many parks and along avenues in countries of mediterranean climate.

Chile still supports some dense woodlands, with a famous fog forest at Fray Jorge, declared a nature reserve in 1941. It is only one of Chile's numerous botanical treasures. The flora of central Chile may include eighteen hundred to two thousand species, with as many as 95 percent native to Chile. This genetic legacy has vital implications for production and conservation.

Africa's Southern Cape

The flora of the Cape region is one of the richest in the world and one of the most seriously endangered. A leading authority on the area, Antony Hall, has estimated that in the Cape and adjacent Karoo, fifteen hundred species are seriously threatened and nearly 40 already extinct. Others suggest that nearly five hundred species are threatened and 60 extinct in the Cape region alone. As in other regions of mediterranean climate, the influence of humanity on a rich and irreplaceable heritage has been little short of catastrophic.

Yet well-developed families of flowering plants, including a thousand species which belong to the daisy family, and many hundreds of bulb plants, succulents, legumes, and others are still to be found. More than eighty-five hundred species of vascular plants are known from the Cape region and nearly three-quarters of these are endemic; such a high proportion of native species is usually associated only with isolated islands. A further puzzlement is that annuals constitute less than 7 percent of the flora. Annuals comprise 27 percent of California's flora, a figure more in line with mediterranean regions of the Old World.

The distinctive Cape flora is usually characterized by its fynbos vegetation—a type of heathland with many colorful species, often of a fine-leaved appearance, which thrive on soil which has lost nutrients through leaching. The broad-leaved matorral, also known as renosterveld, occurs on richer land. Boundaries between fynbos and this scrub are not always distinct, and the two sometimes form a mosaic.

Human influence has been most strongly felt on the fertile areas, where little original vegetation remains. The renosterveld has been reduced from 36 percent of the Cape region to only one percent today, but most of the fynbos on poor soils survives nearly intact. Farmland, urban development, and introduced vegetation cover more than fifty percent of the southwest Cape region.

It seems that tall woody plants and perennial grasses were formerly far more common in the renosterveld. Trees and other large species supplied firewood and construction materials. As in California and elsewhere, many endemic perennial grasses were replaced by annual ones from abroad. The first alien plants were introduced after 1652 when the Dutch East India Company set up a station at the Cape of Good Hope. Particularly serious are problems caused by the large number of woody plants, many of Australian origin. Amongst these are several acacias or wattles, the stinkbean, and three gums of the eucalyptus family. Other woody invaders in-

Chile's unique flora includes the monkey puzzle tree (Araucaria araucana), *right, a conifer now threatened in its temperate habitat, and the vividly colored pouch flower* (Calceolaria), *above, a favorite annual which has frequently been hybridized.*

clude two pines from the Mediterranean Basin.

Many of these invasive species have better powers of regeneration after fire than the native scrub species. Several of these exotics are considered uncontrollable and occupy large stands in many areas. The fynbos have been particularly susceptible, with perhaps 60 percent damaged through competition with the newcomers. When one remembers how much of the native vegetation has been replaced by agriculture and urbanization, it is clear that the flora of the Cape region has been dramatically altered, much of it permanently.

Serious problems are posed by such introductions as *Lantana camara*, one of the world's ten worst weeds. Toxic to livestock, it has invaded the veld and agricultural lands as well as derelict acres. The prickly pear cacti have also become troublesome, with *Opuntia aurantiaca*, able to resist chemical and mechanical attack, presenting the greatest threat.

Australia

This island continent, almost exactly the same in area as the "lower 48" United States, presents a bewilderingly rich flora. Its varied complement of mediterranean plants is in some ways comparable to that of the Cape region. In areas influenced by agriculture and other human activities, little remains of the original ground cover, with only the driest parts remaining wild. Grazing land and fields of wheat occupy the most territory. An authority in the region, R.L. Specht, has observed that modern farming techniques have transformed the vegetation and landscape since the 1930s to an extent that required two thousand years in the Mediterranean Basin.

Australia's zone of mediterranean climate extends from western Victoria state in the east, to the southwestern end of the continent near Perth. As with Africa's Cape region, some rain does fall during summer. Within the zone there is a gradation from humid to semi-arid. Correspondingly, the vegetation ranges from *Eucalyptus* forest to woodland and then to mallee scrub. The term mallee, an aboriginal one, indicates the mode of growth—massive woody tubers from which several stout, tall stems arise—of about one hundred and thirty species.

A fertile coastal strip separating the inland desert of western Australia from the Indian Ocean, the rolling hills of the Darling Range include pastureland for cattle mixed among a variety of woodland settings.

Various eucalypts appear in the open forest of wetter areas and as lesser woodland in drier parts. Where soils are poor, mallee consists of open scrub with an understory of heath. With less moisture, scrub becomes prominent. Overall, semi-arid mallee shrubland forms about 54 percent of the area of Australia's mediterranean territory. Open forests and woodland of eucalyptus are common in moist areas, and shrubby mallee species occur where soils are rocky and poor in nutrients.

The forests are used in many ways, for conservation of water resources, for recreation, and for the production of wood products. Clear-felling and intensive management have been practiced on the better quality forests, and in recent years woodchipping of the various eucalyptus species has increased.

Wind-blown fires, generating intense heat, can severely damage the open forests of eucalyptus and endanger human life. These holocausts occur every three to five years, usually during summer and autumn. Denser stands of mallee scrub may succumb to fire at intervals of fifteen years to a century, depending on the density of the underbrush and the amount of combustible wood, leaves, and other litter. The flammable oils and resins in the eucalypts enhance the risk of fire, yet most species are adapted to fire conditions and have various strategies for renewal.

As noted earlier, new trees can emerge from the woody tubers. It is a saving grace.

Conclusion

It may seem troublesome for the reader, at first, to encounter repetitions of the great genetic potential of native species that somehow manage to survive agriculture, urbanization, and other development and activity. Parts of Australia fit the pattern, too. Here as elsewhere nature's powers of regeneration are immensely important for the enduring benefit of people of the various lands of mediterranean climate.

Here and there, people are beginning to learn how to move beyond the conventional uses of the green world's priceless genetic bounty. In Australia, for instance, development of domesticated forage crops from local wild plants has begun an agricultural revolution whose successes can be repeated. Experience gained in any one region may have applications in them all. Each has its own underexploited plants which can be turned into cash and also help to protect and regenerate soil and water resources. Despite great loss of productivity over recent centuries and the large scale destruction of habitat, the lands of Mediterranean climate still have much to offer if we learn how to recognize underutilized botanical resources and use them to their best advantage.

Numerous plant species of southwestern Australia are now threatened. Jacksonia furcellata, *above right, of the pea family, and* Eremaea beaufortioides, *above left, of the myrtle family, are not yet in grave danger, but flower exports have imperiled* Anigozanthus flavidus, *"yellow kangaroo paws," left.*

Temperate
Splendor

IN GREAT WOODLANDS OF THE temperate zones, exemplified here by a coniferous forest near Jackson Hole in the Grand Tetons of Wyoming, only a few species comprise the dominant vegetation, and give it a rather uniform aspect. Worldwide, the variety of temperate forest trees is provided by some 100 species of pine (*Pinus*), 200 maples (*Acer*), 450 species of oak (*Quercus*) and 500 of willow (*Salix*), although usually only two to four of the species may mix in any given forest. For example, in the world's largest forest, the coniferous "taiga" of the Soviet Union, an estimated 38 percent of the cover consists of 2.7 billion acres of Siberian larch, more than 4 million square miles. By contrast, tropical forests are comprised of many tree species and quite widely scattered populations as well as individual trees.

In the low latitudes, i.e. those nearer to the tropics, the temperate zone forests consist mostly of deciduous trees that seasonally drop their leaves at the onset of cold weather. In the higher, more northerly latitudes, evergreen conifers predominate. Their needle-like leaves, covered with a waxy layer, allow little water to escape from the tree, especially in cold, dry winters.

Soils of the temperate regions vary more widely than do those of the tropics—especially those of the high forests of the moist tropics where rain has washed away nutrients during thousands of years. In temperate zones, however, organic matter has a greater opportunity to accumulate and enrich

The radiant globeflower (Trollius altaicus), *above, flourishes in the alpine meadows of Mt. Suptun, left, visited by a joint (1983) Soviet-United States botanical expedition to Altai mountains of Siberia. Top, bouquet at a campsite in the remote Sayan mountain range of the Tuvin Autonomous Republic.*

The ten trees shown below in their relative scale represent the largest standing individuals of various species in North America, recently recorded by the American Forestry Association. The four conifers (Giant Sequoia, Douglas Fir, Ponderosa Pine, and Southern Cypress) are of far more ancient lineage than the six flowering, broadleaf, and hardwood trees depicted.

Giant Sequoia
272'

Douglas Fir
221'

Ponderosa Pine
162'

White Birch
96'

American Elm
160'

The twin seeds of the pinyon pine (*Pinus monophylla* or *Pinus edulis*), a conifer from the southwestern United States and Mexico, are borne "naked" on cone scales and exposed to the wind for pollination. Left to right, the embryo splits its seed coat and the root penetrates the soil, from which it absorbs water and minerals. As the ten green cotyledons (seed leaves) emerge in a whorl from the seed casing, they begin to convert sunlight into food through photosynthesis. This process in turn fuels the growth of the plant's true leaves from the terminal vegetative bud at the center of the whorl.

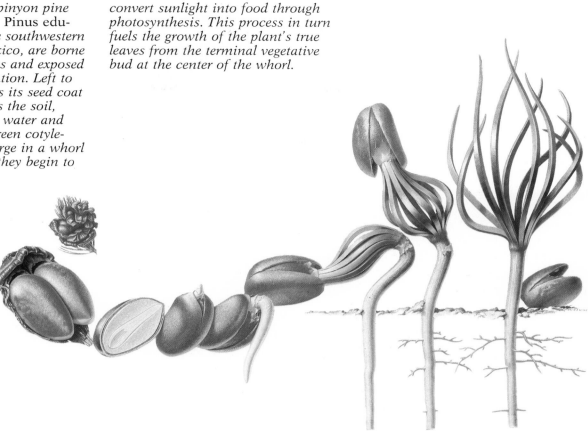

Oak
8'

Sugar Maple
116'

Black Cherry
102'

Southern Cypress
122'

Shellbark Hickory
122'

Morning mist in Nova Scotia, enhances the rich and redolent splendor of the northern forest, below. Leaf of wild black cherry, a member of the rose family, adds its brilliance to the landscape, right. Wisp of web clings to a dewy bud in Virginia.

the soil. Prehistoric marshes associated with continental glaciation accumulated large amounts of peat. When the glaciers retreated, the peat remained—to the benefit of farmers today. Temperate zone soils generally have higher levels of organic matter than jungle soils. Therefore they can take higher levels of mistreatment and still continue producing.

The foundation of modern soil science was laid in the Soviet Union—an area lying mostly in the north temperate region, the domain of the great north woods. Many Russian words describing temperate soils and their characteristics have been adopted by scholars.

Below; a Lycopodium, *or running ground pine, is no pine at all but a club moss of very ancient lineage. In Carboniferous times, its kind— though giant then—helped to form earth's great deposits of coal. Today the increasingly rare plant is sold as Christmas greenery. It is most abundant in Canada and the northern tier of the United States. Heavy headed member of the buttercup family,* Pulsatilla rubra, *exhibits hairy buds and stems. It grows in the Auvergne, highlands of France's Massif Central. Opposite; coastal redwoods loom in California fog.*

Tropical Rain Forests

THE TROPICAL RAIN FOREST HAS LONG BEEN a source of mystery and wonderment, especially for those who know it only from pictures or report. The dense steaming jungle, replete with exotic birds and insects, conjures up visions of nineteenth-century naturalists such as Alexander von Humboldt and Charles Darwin, revealing to a receptive Victorian society the treasures and challenges of the great unexplored tropics. Avid readers traveled by armchair with the likes of Thomas Belt on his adventures in the wilderness of Nicaragua:

> . . . we entered the primeval forest. On each side of the road great trees towered up, carrying their crowns out of sight amongst a canopy of foliage; lianas wound round every trunk and hung from every bough, passing from tree to tree, and entangling the giants in a great network of coiling cables, as the serpents did Laocoon; the simile being strengthened by the fact that many of the trees are really strangled in the winding folds. Sometimes a tree appears covered with beautiful flowers, which do not belong to it, but to one of the lianas that twines through its branches and sends down great rope-like stems to the ground. Climbing ferns and vanilla cling to the trunks, and a thousand epiphytes perch themselves on the branches. Amongst these are large arums that send down aerial roots, tough and strong, and universally used instead of cordage by the natives. Amongst the undergrowth several small species of palms, varying in height from two to fifteen feet, are common; and now and then magnificent tree ferns, sending off their feathery crowns twenty feet from the ground, delight the sight with their graceful elegance. Great broad-leaved heliconiae, leathery melastomae, and succulent-stemmed, lop-sided leaved begonias are abundant, and typical of tropical American forests.

The attitude of society in developed countries towards jungles has been one of admiration rather than comprehension, largely because the jungles are so remote. The vast forests of the Amazon, Southeast Asia, and Africa have always been there, and it has been assumed that they always will be.

Hyla geographica *stalks insects in Brazilian Amazonia, near Manaus. Trunks and branches of rain forest trees provide the habitat for epiphytic ferns, as on Borneo's Mt. Kinabalu, opposite.*

Overleaf; Brazilian Indians reveal the girth of a plank-buttressed, emergent-layer tree to the botanist Karl von Martius and the zoologist Johann von Spix during the renowned Bavarian explorers' 1819 expedition through the Amazon Basin; see also page 211.

Biological Riches

There are many kinds of tropical rain forest, depending on climate, especially rainfall, the underlying substrate, and altitude. What makes all of them so interesting, so valuable, is their richness in species, compared with forests of the temperate zones. The jungle's complex, multi-storied structure provides delight and data for scientists.

These are the most highly developed ecosystems on earth, taking shape over thousands of years and producing large numbers of diverse and unusual species which have evolved in and with them. In a very real sense these forests are our richest gene pool, the largest and least explored source of new natural products. The plants, mammals, insects, and birds which they contained in such diversity have intrigued and perplexed biologists for centuries. Yet very little is known about the detailed composition, structure, and functioning of such forests, and we are equally ignorant of their silviculture and proper management. A further complication: the need for economic development is greatest in those areas of tropical forest with most to lose from unwise exploitation. Yet we are still very uncertain as to the most effective means of combining development of such forests with conservation of their resources.

Much of the world's tropical rain forest has already been lost and, as we shall see, it is today being destroyed or converted to other use on a vast scale. If we had been able to photograph the Earth from a satellite a century and a half ago, we would have detected on its land masses a band of green extending some ten degrees north and south of the Equator. These tropical rain forests and monsoon forests, partly covered with cloud, occupied some six million square miles, about twice the area of continental Eu-

rope. It is reckoned that 3.5 million square miles of the forest are left today—their overall extent having been reduced by something approaching a half since the 1830s.

India, Sri Lanka, and Burma have suffered the greatest loss, a staggering two-thirds of the tropical forest during historic time. Africa has lost half. The forests of Southeast Asia and Latin America have been reduced by 35 and 40 percent, respectively. Even so, the tropical moist forests represent a major part of the world's vegetation, and any substantial changes to them, as to other major plant areas, must be a matter of concern to us all.

Climate and Soil—Variations on a Theme

The richest and most resplendent jungles thrive on soils rich in clay, with six feet or more of annual rainfall distributed evenly throughout the year—ideal conditions for uninterrupted

plant growth. These favored regions also contain the largest number of commercially valued tree species and, being relatively accessible, face the greatest risk of overexploitation and destruction of habitat. Forests of this sort occur in the Amazon Basin, central Africa, and Southeast Asia. Swampy and seaboard areas often provide anchor for the mangroves, an important life zone covered in the next chapter.

Further inland, inundated areas develop swamp forests of various kinds. Little humus builds up in them, so peat is scarce. Where the water is poor in minerals, however, some peat is occasionally formed. In these areas—Southeast Asia including Sumatra, Malaysia, and here and there in equatorial America and Africa, evergreens usually dominate. As in other types of swamp forest, the number of species is restricted, and knee roots and other natural obstacles restrict access.

Where poor but well-drained soils such as coarse sand acquire a hardpan of iron minerals, a distinctive type of lowland heath forest emerges. This dense formation of relatively small-leaved trees of low stature and uniform canopy is exemplified by Brazil's caatinga and campos.

The kinds of forest that develop on slopes and in mountain valleys contain fewer species than those of luxuriant lowland growth. Trees are shorter in such sub-montane forest, with only two stories instead of three, lower canopies, smaller buttresses, many epiphytes (air plants), along with mosses, liverworts, and hardly any vines. Higher still, the upper montane or cloud forest unfolds. Plants drip with water condensed from fog and clouds, and small leathery-leaved trees form an even canopy. As in lower forest zones, epiphytic plants, liverworts, filmy ferns and mosses abound, giving rise to the term "mossy forest."

Above the cloud forest there may develop an elfin woodland of low stature trees. The sub-alpine forest, inching up to timber line, contains ever more densely packed trees with smaller and

Mist cloaks Sugar Loaf and the harbor of Rio de Janeiro. Even as Brazil's coastal cities grow, settlers are moving more deeply into moist, tropical Amazonia where even fence poles sprout. Peru, Ecuador, Colombia, Bolivia, and Venezuela also have Amazonian territory.

smaller leaves. Finally, vascular plants disappear and only lichens and algae survive.

Layers of the Rain Forest

The tops of the different sized trees and shrubs typically form as many as five layers, or stories. We usually discern three main tree layers plus the understory of shrubs and herbs which may itself have two layers or three.

The uppermost or emergent layer consists of trees of 100 feet or more in height. With their tops rarely touching, these giants form a broken canopy. Lower down and more continuous, the main canopy is composed of the tops of closely spaced trees 50 to 100 feet high with broad or rounded crowns. This formation, pierced by the emergent giants, creates a "broccoli-tops" appearance, an effect of light playing on treetops at different heights. The lower story of smaller trees, some up to 50 feet high and with rounded or elongate crowns, forms a more or less dense layer just below the crowns of the upper story.

Each level has its own microclimate. Tree tops in the emergent layer bask in full sunlight and are subjected to high temperatures and low humidity. They also absorb the force of strong winds. Lower down, the humidity increases and the amount of rainfall and sunlight penetrating the canopy decreases.

At ground level the air is very still, the light very dim. Investigation suggests that only two percent of the sunlight at the exposed canopy penetrates to the forest floor. Flecks of sunshine may touch the litter of fallen leaves, seeds, and fungus on the ground, but the rest is a diffuse and reflected greenish glimmer.

One might surmise from its richness above ground that the rain forest arises on highly fertile ground, but the reverse is usually true. Most jungle earth is very old and deeply weathered by thousands of years of rainfall. The relentless splatter has washed out the nutrients, leaving little plant food and only pockets of humus. The hard residue is rich only in iron and aluminum, both in the form of insoluble oxides which combine to give skeletal tropical soils—known as laterites—their characteristic red or yellow tones. While jungle soils are quite deep, even up to two hundred feet of weathered rock, the surface layer in which most of the roots develop is thin and fragile. Sometimes it is white sand.

How can we reconcile the jungle's lavish growth with its soil's dearth of nutrients? The answer to the paradox lies in the fact that such

The warmth and humidity of tropical America support over eighty-two hundred orchid species—more than any other region of the world. The four pictured here are all epiphytic on the bark of trees. Opposite left, this mitten-shaped orchid is only one of over a thousand Pleurothallis *species that flourish in Latin America. The Brazilian* Miltonia clowesii, *opposite center, is one of 25 American members of its genus. Opposite right is* Oncidium nanum, *native to Brazil and Peru. At right, the showy* Masdevallia veitchiana, *also from Peru, is one of more than 300* Masdevallia *species. Brazil and Argentina have both established national parks adjoining the Iguaçu River at Iguaçu Falls, below, along their common border. With a crestline over 2.5 miles long, the cataract is longer and wider than Niagara Falls; its misty spray promotes luxuriant vegetative growth in the surrounding area.*

a forest is, in effect, a closed and leak-proof system in which the nutrients circulate rapidly. Fertility resides in the vegetation, not in the impoverished soil.

As the American ecologist Carl Jordan has commented:

> *So well does the living forest hold onto the scarce nutrients that nothing can release them, except the destruction of the forest itself.*

Though tropical trees remain green all year, they constantly shed leaves. This litter hardly ever has a chance to accumulate: insects and fungus make short work of it, thus aiding the rapid recycling of nutrients back into the vegetation. In contrast, temperate forests often possess a deep and fertile layer of topsoil and leaf litter that may take years to decompose.

It is for this very reason that clearing or conversion of the forest to other use usually leads to the nutrients being wasted, by burning, by removal of trees and other plants, or by decomposition of the thin layer of humus or litter on the forest floor. Once the trees have gone, the soil layer can be blown away or eroded by water.

It is against this background that both utilization and conversion of the forest for other use (often a polite way of saying destruction) has to be considered. On the one hand, the high productivity of the rain forest all but evaporates in the conversion process. On the other, substantial loss of nutrients can seriously impede forest recovery and regeneration. Such considerations have led the Mexican botanist Arturo Gómez-Pompa to suggest that many tropical rain forests must be regarded as a non-renewable resource. They also make it obvious that large-scale clearing of the Amazon rain forest would be unwise since most soils are simply not suited to the intensive, mechanized agriculture practiced in the temperate regions of richer and more resilient soils.

Direct comparison of rich soils with poor ones in various forests suggests reasons for this pessimistic conclusion. One such study involved four areas: two temperate forests; one rain forest with rich soil in Puerto Rico's highlands; and a site on typically poor Amazonian laterite. Peak production of wood and foliage in each was comparable despite the obvious seasonal differences and variations in soil nutrients. With the Amazonian forest, though, it became clear that the key to survival lay in a thin but rich layer of roots and humus at the soil's surface.

Investigations have indicated that the poorer the soil the thicker the root mat. This network with its fine root hairs is remarkably effective in catching and holding onto nutrients, a process called adsorption. In one experiment in the Amazonian jungle, a sample of calcium and phosphate sprinkled on a very thin root mat resulted in the immediate adsorption of up to 99.9 percent of the plant food materials.

The root mat acts as a means of transferring nutrients from decomposing leaves to the roots by the action of microscopic fungi called mycorrhiza, which break down fiber, digest and assimilate part of it, and excrete simple chemicals which the trees can consume. In a sense, the mycorrhiza act as a nutritional bridge.

Another mechanism: the root mat is acidic due to the presence of tannins. These chemicals largely prevent the growth of bacteria which might cause a loss of nitrogen. In fact, virtually no nitrogen consuming bacteria could be detected on some rain forest soils in Venezuela.

Adsorption also occurs high in the canopy where algae and lichens cling to the leaves and catch and hold some nutrients dissolved in rainwater. In this case chemical elements acquired by filtering of particles from the air and the fixing of nitrogen through lightning discharge are selectively removed as the water trickles down the surfaces of living leaves, some of which have evolved enlongated drip tips.

By its many conservation mechanisms, the rain forest is adapted to the maximum extraction of all nutrients that become available to it. The poorer the soil, the more the plant must depend on ecological mechanisms. These are not physical or chemical components of the soil, but they are nontheless integral to the living structure of the rain forest. Nearly all the rain forest's strength is tied up in living root, trunk, branch, stem, fiber, and leaf. Destruction of the forest short-circuits the nutrient conserving mechanisms and allows the fertility to bleed away.

Myriad Species

Few species are found in temperate forests, but very many individual trees of one or a few species grow within a given area, much as the well-known oak woods, beech forests, ash groves, and the conifers of many kinds. Tropical rain forests may contain as many as a hundred different species per acre, but relatively few trees of any one species in a single area. Animals are also scattered about in the tropics. In the case of plants, heavy exploitation of any one species can lead to its local extinction, since large and heavy jungle seeds seldom travel far from the parent trees. In contrast, temperate zone trees more often drop tiny seeds with wings or parachutes for wind dispersal, or which snag in the fur or feathers of animals.

Furthermore, there is an inverse relationship between the distribution of botanists and the distribution of plant species. In most South American countries we find a mere handful of botanists. In an extreme case such as Bolivia, there is only one native-born botanist. By far the largest flora is that of tropical America, where nearly one hundred thousand species dwarf the thirty-five thousand of tropical Asia and the thirty-thousand of Africa south of the Sahara. Globally, of the quarter million species of flowering plants, one hundred and sixty thousand

are tropical. The flora of Brazil alone is estimated at fifty-five to eighty-five thousand species, with Colombia a close second. The true magnitude of such figures becomes evident when one compares them with the total flora of the United States; fifteen thousand species, or continental Europe's twelve thousand or fewer! While a professional taxonomist could come to know and recognize as many as fifteen thousand species during his or her lifetime, coming to grips with fifty thousand species is quite out of the question.

A phalanx of caterpillars eats its way across a leaf. Fungi and bacteria will follow in its wake. Right, protective coloration allows a rainforest moth to become "invisible" to predators when it settles on a showy Heliconia, *related to the bird-of-paradise plants.*

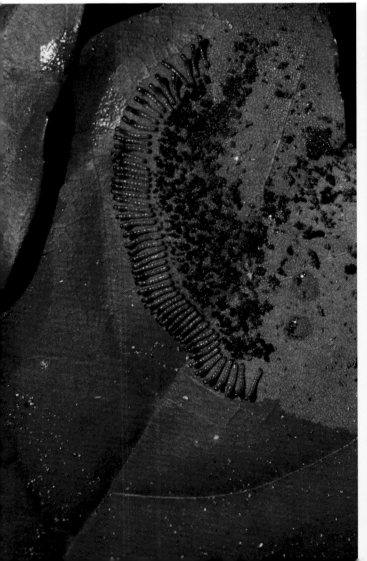

The concentration of ferns in tropical regions is even more dramatic, with eleven thousand out of the planet's twelve thousand species appearing only in the tropics. The situation for the fungi is very disturbing since probably ninety thousand species out of an estimated world total of one hundred and twenty thousand are only tropical. Although there has never been anything approaching a complete catalogue of species for any major tropical flora or fauna, an even greater diversity appears in animal groups. It has been estimated that of the two million species of insects believed to live in the world today, about three-quarters of them are tropical. It is alarming that most of them have yet to be described and that about half of them could well become extinct if destruction of the tropical habitat continues at its present rate. This is likely to happen—and before botanists and entomologists can possibly discover and describe either the plants or their insect pollinators.

Stability Versus Change

Because of their sheer scale and complexity, rain forests give the impression of being permanent and static, but this is largely an illusion. When ecologists describe the forest as being stable, they refer to its ability to continually adapt, to adjust year by year. Close observation reveals a complex pattern of renewal of species and populations. The forest is thus able to respond to minor climatic changes and varying biotic pressures with only slight alterations in species composition, distribution, and overall structure.

Although the mass equivalent of the living forest may die and be renewed once or twice a century, it would appear to be the same forest to an observer who viewed it every 25 or 50 years during a period of five hundred years. It depends, of course, on the size of the forest concerned but, by and large, the tropical rain forest is a system that maintains itself during hundreds and thousands of years. We have gained this knowledge from various lines of paleo-geographical and paleo-climatic evidence.

One should therefore look upon the tropical forest as an extremely rich and complex community of plants and animals capable of surviving largely unchanged for perhaps thousands of years. Yet the jungle is totally dependent for its survival on the absence of interference with its finely tuned system of nutrient recycling.

In such a delicately balanced system, each alteration—by the action of nature (such as the death or collapse of a tree) or of human beings (as by logging too many trees)—disturbs the ecological equilibrium and sets a series of changes in motion. Just how serious these changes are depends on the degree of disruption to the nutrient cycle. Furthermore, in denuded areas, the soil will certainly be exposed to higher temperatures than before. Also, the incidence of light and rainfall are altered and, in turn, such factors affect the amount and kind of regeneration. If the disturbance is drastic, the consequences can be serious and irreversible. Yet some shifts may be beneficial, a part of the natural order.

As scientists we are all too aware of the limits of our knowledge. For instance, the reproductive strategies of the rain forest trees and other species are seldom known in any detail, if at all. Very few studies, in fact, have been made on the times of flowering, fruiting, viability of seed, germination, pollination, and so on. Some of the species come into bloom more or less continuously, others sporadically or seasonally; or they may vary from year to year. Some bamboos flower only a few times a century or indeed only once in more than a century.

The factors responsible for such variations are little known. The various factors also interact, and the interactions themselves will depend, in one degree or another, on how far the pattern of flowering, fruiting, and so on is inherited and how far it is controlled by the environment. We need to know just how these and other factors express themselves in the changing world of the rain forest. Not without exaggeration have tropical rain forests been described as one of the world's great outdoor laboratories in which the forces of evolution are deployed to the full.

Because we are so largely ignorant of the detailed ways in which the tropical rain forest ecosystem functions, we cannot predict with accuracy how the forest will respond to particular alterations. We do know that when a forest is destroyed by fire or by clearing for agriculture, the nutrient balance is upset and the loss of forest cover can be permanent. Nutrients, suddenly released from the organic matter and the ash, are ready to be washed away.

Pristine forests are often described as virgin, primary, old growth, or climax. Those obviously tampered with are secondary (or degraded). Even so, virgin forests are seldom untouched since most tropical rainforests, except in inaccessible areas, have been altered by the people

Swinging through the lianas, this young orang-utan covers large areas of the Borneo rain forest in a search for food. Each night, Pongo pygmaeus builds its nest of branches lined with leaves. Apparently as nimble of mind as they are of limb, these rather close relations of humans (orang-utan means "jungle man") seem to remember the sites of choice food sources and return at times when fruit is ripe. Though rarely hunted, the orangs are severely threatened by progressive destruction of their habitat for agricultural land and timber; their population has been reduced to about five thousand. Fire followed drought in the early 1980s and further damaged the rain forest of Borneo. As their green life line is gradually severed, the orang-utans move closer to extinction.

who live in them and off them. One estimate places 200 million people living within or at the fringes of forests. Over the centuries, many such inhabitants have developed a special relationship with their native woodlands, a feeling for the life of the forest. Such may be artistic, in a sense, arising as it does from an intimate association with such a complex environment, with all its subtle changes and varied and varying reactions. In spite of the infinite nuances of the forest, these people have learned to live in harmony with it, and without destroying it.

Use and Misuse of the Forest

While tropical rain forests have been exploited by human beings for many centuries, no special problems arose as long as the woodland's capacity for recuperation was not exceeded. Local people encountered few problems of management so long as they used the land on a shifting or subsistence basis, clearing only enough to farm for a season or two before they moved on. Such slash-and-burn cultivation, or shifting agriculture, is often named as a major cause of forest destruction. In many circumstances, though, it has been an efficient method of land management in areas of low population density.

Shifting agriculture is the process of slashing a clearing in the forest, burning much of the fallen vegetation, and cultivating a mixture of crops for two or three years until the soil loses its productivity. The transient farmer then repeats the process elsewhere and the formerly cultivated ground is allowed to regenerate itself. While the area of forest cleared by shifting cultivators is enormous, perhaps as much as seventy-five thousand square miles annually, the practice is by no means catastrophic, seldom as damaging as generally thought unless done across huge areas of land.

As it has evolved over the centuries in Africa, Asia, and Latin America, the system of shifting agriculture allows for fallow periods of ten or even twenty years. Soil fertility can be restored with gradual development of a secondary succession of trees on the formerly cultivated land. The secondary forest is later cleared for cultivation and the process is repeated. Older land is often most suitable for use because the secondary forest holds softer trees and the farmer works less the second time around. Indeed many forest farmers exploit only secondary forest and thus prevent its ultimate regeneration.

Some of the modern agroforestry systems embody this principle of rotation in an attempt to transform slash-and-burn into new and sustainable forms of continuous land use. Shifting agriculture may be more labor intensive than some conventional systems but, in general, it is not a profitable use of large forested areas. More permanent agriculture would, however, often require considerable expenditure on fertilizers and pesticides and would require more labor. Thus we have one of those vicious cycles which tend to keep many farmers of the developing world from achieving sustainable prosperity.

Poverty's hold can be broken, however, but with greater difficulty as the density of human population increases. In this case, shifting agriculture becomes a serious problem and a threat to the world's tropical rain forests. Slash-and-burn employed on a large scale jeopardizes the former balance between nature's capacity to produce and our ability to exploit. The same problem occurs when farmers practice slash-and-burn in the wake of commercial loggers whose roads provide access to formerly inaccessible regions. The logging itself may represent the least of the destruction. Traditional inhabitants who know the forest are now joined by large numbers of immigrants. Settlers not only cultivate the areas cleared by the loggers, but cut down and burn more trees to gain the ash needed to fertilize their crops.

Unable to find employment or to obtain agricultural land elsewhere, the newcomers may have been directed to forest areas by governments which regard the forests as a commercial resource suitable for any and all exploitation. Lacking the necessary expertise and traditions, these immigrants seldom allow for long fallow periods. Soon the land is capable of sustaining neither crops nor trees. People move on to damage other areas of the forest or become ecological refugees of the kind who also emerge from overworked grasslands and other regions of desertification.

It is difficult to estimate with any accuracy the amount of conversion of tropical rain forests. The Food and Agricultural Organization of the United Nations (FAO) goes to great pains to prepare accurate figures. An accessible and reliable guide is the study *Conversion of Tropical Moist Forests* commissioned by the National Research Council of the United States Committee on Research Priorities on Tropical Biology. In it, Norman Myers has concluded that forest farming could well account for the conversion of at least thirty-five thousand square miles of primary forest to permanent cultivation each year. This does not include figures for the exploitation of secondary forests.

The greatest losses may have been suffered in Southeast Asia, with particular danger to Indonesia from shifting cultivation. Tropical Africa is much more mixed, but perhaps a million square miles of tropical rain forest were converted to agriculture by shifting cultivators before modern patterns of economic development were imposed after Independence. In central African countries such as Gabon and Equatorial Guinea, low population density may allow for shifting cultivation without too great an effect on the quality and composition of the primary forest. Regeneration may be possible.

Latin American land is subjected to great stress. While the governments of Brazil and Colombia have sponsored colonization and settlement programs, the Trans-Amazon highway has channeled waves of migration from the overpopulated northeast into the heart of the Amazon Basin. Colonists expected rich soils for high yields but soon found that Amazonian laterite was unable to support crops for more than a few years. Their land's brief spurt of productivity was predicated on the burning of vegetation which opened the soil to erosion and the loss of nutrients.

Cattle Ranching

The conversion of tropical rain forests to pasture for cattle is one of the most pernicious and wasteful uses yet devised. It seldom benefits the peoples of the countries concerned, though it has made handsome profits for large-scale entre-

Hot springs rise at Cidodja in Java's jungle. At more than 170° F, these waters support tenacious blue-green algae, among the most primitive and most ancient members of the Plant World.

quently. Then, hooves of grazing cattle pack the ground and lead to soil erosion.

With the land exhausted after five or ten years, production dips sharply. Ranchers abandon the land to invading weeds and scrub, and move on in the manner of the forest farmer to cut down other areas of forest and start the cycle off again. The incentive for much of this peculiar brand of agribusiness comes from the strongly developed markets of the United States and western Europe, where any source of cheap beef gives a great economic advantage to the vendor. Some Latin American governments have encouraged beef production by means of subsidies, tax incentives, and other forms of aid. Since this beef-export business depends on the conversion of tropical forest, the cheap hamburger is in this very direct way a major cause of deforestation in the humid tropics.

It has also been suggested that this type of cattle raising stems from a strong political compulsion by governments to assert sovereignty over the vast and so far little developed areas of rain forest. Since developed countries of the temperate zones often began their growth by cutting down forests, this is hardly surprising. Whatever the reasons, it is becoming clear that insufficient attention has been paid to the factors which determine the productivity of the forests and to the consequences of this assault. Huge tracts of land are cleared, bringing temporary gain and then loss—probably permanent.

Unfair as it may seem, temperate zone approaches will ruin most tropical moist forests. Desire for profit can be creative and commendable, but not as practiced in many rain forest areas. It is not as if the people profit widely. Much of the benefit is gained by multi-national enterprises which became involved in these huge ranching ventures—many of which have ultimately failed as Tom Lovejoy of the World Wildlife Fund has reported.

Productivity of cattle companies in Brazil is much lower than had been hoped for; several such operations have already been abandoned because of low profitability and perhaps two

preneurs, often government-sponsored, and has benefitted the consumer of cheap beef in the United States through the so-called hamburger connection. Some of this meat, though much less, makes its way to Europe.

In America's tropics, beef production has involved converting rain forest into pasture as quickly and cheaply as possible, and on a vast scale. The trees are generally cut and burned, even though the marketable timber could fetch a price. Calculations suggest that an average of 30 cubic yards of useful timber goes up in smoke for each acre burned; the loss of revenue to date is on the order of $10 billion.

There is every reason to believe that the soil will go into permanent decline after only a few years of grazing. It is difficult to maintain productivity without the application of expensive and often unavailable fertilizers and pesticides. The practical solution is to burn the pasture fre-

hundred more are likely to follow. Now Colombia and Peru may be adopting Brazil's example—disasters and all—in their own Amazonian territories. In some cases, such plans have been supported by international agencies. The establishment of cattle ranches sponsored by Brazil's Amazon Development Authority (SUDAM) alone has led to the conversion of thirty-thousand square miles of Amazon forests between 1966 and 1978, not to mention clearing on the thousands of ranches established by other means.

Central America's situation is far more critical. There is an old tradition of cattle raising for export. The beef goes mainly to the United States, where the cost of raising a cow is two to four times higher. The convenience food industry turns most of this meat into hamburgers and frankfurters.

Like all such factors, it is difficult to calculate with any precision the overall effect of cattle ranching on the conversion of the tropical rain forests of Latin America. But the best guesstimate indicates the loss of more than seventy-five hundred square miles a year.

Tropical Timber

Commercial extraction of timber is the main use of the tropical rain forest today. The manner in which the wood is taken represents a major threat to the green world of the tropics. The scale of exploitation is far too great to be sustained indefinitely. With careful and well-considered harvesting, the tropical rain forests could be conserved and developed for the long term by the people who live there. These people can create enduring prosperity for themselves and their countries. This is all too rare today. We note, for instance, that the leguminous tree *Pericopsis elata* or afrormosia was introduced into world trade in the 1950s. This highly esteemed hardwood from tropical Africa has been widely accepted by the furniture and veneer trade. Now, after less than 40 years in commerce, it faces extinction due to heavy logging and poor natural regeneration.

The demand for wood is on the increase throughout the world. This is hardly surprising, considering the supreme versatility and numerous applications of wood. Timber goes into construction beams, plywood, and veneers for furniture. Pulpwood is a source of cellulose in newsprint and other papers, and in the production of chemicals, fibers, synthetic foods, alcohol, and many other products. The global demand for wood is beginning to outstrip the capacity of forests to produce it. Estimates produced by the FAO indicate that overall demand for wood is expected to more than double by century's end.

In particular, the consumption of hardwoods by the developed countries has increased markedly since World War II—and the bulk of the world's hardwoods grow in the tropics. Ninety percent of jungle trees are hardwood, in temperate forests most are coniferous softwoods. The temperate hardwood forests have been steadily depleted over the centuries and those that remain are coming increasingly under protection as the result of forceful campaigns from environmental and conservation groups. In con-

Calathea makoyana gleams in Brazilian sunlight. In the same family as arrowroot, it is a close relative of the florist's prayer plant.

sequence, the demand for hardwoods is being directed more and more to the tropics, often by the very nations which have restricted the cutting of timber in their home territory.

Since 1950, the importation of hardwood timber by the developed countries has increased dramatically: shooting up fifteen-fold. Its use has only doubled in the tropical countries. What is more, a single nation, Japan, is responsible for more than half the developed world's imports of hardwoods from the tropics, the bulk of it coming from the forests of Southeast Asia. As a consequence, reserves in this quarter are rapidly nearing depletion and Japan is now looking to Latin America as a new source of supply.

The United States, the world's second largest importer of tropical hardwood, buys mostly from Southeast and Southern Asia. Eighty percent of hardwood imports arrive on American soil in the form of plywood and veneer that has been either processed or transported *via* Japan, one of the world's single greatest wood processors. Europe—mainly the western nations—imports about a third of all tropical hardwoods, mostly from forests in western and central Africa although imports from Southeast Asia have increased in recent years.

Viewed globally, Southeast Asia accounts for three-quarters of the world's trade in tropical hardwoods and tropical Africa for just about a quarter. Although it harvests ten percent of the world's total, Latin America exports very little timber. This pattern is expected to change over the next 20 years as the forests in Southeast Asia become exhausted. The increased development of reserves in Amazonia and of central Africa, coupled with improved technology, can easily lead to a substantial increase in the exploitation of hardwoods in these two areas. In 1980, international trade in tropical hardwood timber totaled in excess of $8 billion—more than cotton, rubber, and cocoa, and surpassed only by coffee at $12.2 billion. Some African countries depend, for the moment, on timber exports as a major source of foreign exchange.

Tropical hardwoods are big business and present a temptation to governments seeking quick revenue. They often pay little heed to ultimate consequences. As a result, logging concessions have been let to timber companies on a vast scale in many parts of the world. In the Philippines, for example, all accessible dipterocarp forests (teak and other hardwoods) will have been logged out or converted to other non-forestry uses before 1990. In Indonesia, timber concessions for practically all the accessible lowland forests, including some which have protected status, have been let to loggers.

Exploitation of commercially valuable rain forest in the Ivory Coast since the late 1950s has been so "successful" that all exploitable areas could be exhausted by the end of the century. It should be noted, however, that the Ivory Coast may be undergoing a change of heart, prompted by recognition of the seriousness of its environmental problems. A further indication of Ivory Coast's determination to be responsible toward future generations is the establishment of a facility of the United Nations University.

The international paper trade looks increasingly to tropical hardwoods for paper pulp. Although forests of hardwood contain only small proportions of the native softwoods formerly used for pulp, recent technology has made it possible to pulverize wood chips from a variety of hardwoods. Future demand for paper pulp will in part be met from tropical hardwoods.

Forest Exploitation: Old and New Methods

An old-fashioned approach called selective logging not only helped to create the traditional woodworking industries, but permitted conservation of some prized timber species. Single trees of one or two particular species, and of a minimum diameter, were felled by axe and dragged to the edge of the forest or to the nearest river. Logs were carted off or floated to market. This method produced few and relatively small gaps in the rain forest canopy. As we have learned, tropical tree species are widely scattered so very few logs of any one species were harvested per acre. Slow-growing, shade-loving species eventually healed the break. The net result was a temporary depletion in the number of certain species of certain sizes but hardly a disaster.

It would be wrong, however, to suggest that selective logging was always so well controlled and ecologically innocuous. Indeed, examples abound of excessive exploitation of rain forests. We have the case of Burma in the nineteenth century. In his study of historical ecology, *King Thebau and the Ecological Rape of Burma*, C.L. Keeton has argued that the Dry-Zone drought of 1883 to 1885 was largely a result of excessive deforestation by the French and the British. Agents of distant commercial interests vied with each other in the race to exploit teak and cutch (*Acacia catechu*) in the provinces to the south of

Mandalay. And of course massive deforestation was a necessary precursor of the introduction of plantation crops such as tea, coffee, and rubber in Southeast Asia and elsewhere.

Selective logging is still practiced today, but on a much greater scale and with serious implications for the continued existence of the forest. Despite the enormous diversity of hardwood species, few meet with acceptance in the marketplace. Traditional practices of forest management and exploitation have a bearing, as does the very conservative nature of the international timber trade. Those involved show a distinct reluctance to handle more than a small number of the available species. Such selective harvesting has been called a "creaming" operation. Statistics from FAO indicate that of the one hundred and five commercial species in western and central Africa, only thirty-three species or so make up the bulk of exports. More than six hundred and thirty species are available in Southeast Asia, but only a dozen comprise the main exports. Tropical America produces two hundred and twelve species suitable for the hardwood trade.

Selective logging as currently undertaken in the tropics can be very wasteful as heavy equipment can severely damage trees and seedlings that remain on the logging site. The ripple effect that results is not surprising in light of the complex three-dimensional structure of the rain forest. The canopy of tree crowns is often so intricately linked with lianas that the felling of a broad-crowned tree brings down or breaks off parts of its neighbors, along with all the attached climbers and epiphytes. Animal inhabitants of the forest also suffer.

Furthermore, moving harvested trunks through the forest can also cause serious damage. In one operation in Nigeria, the felling of three trees per acre, with a yield of 25 cubic yards, injured up to 32 percent of the remaining trees. In some forests of Malaysia 10 percent of the forest was cut, but half of the trees in the area of operations sustained damage as sawn trunks crashed to earth or were dragged from the site. Diseases can infect broken branches and also spread to nearby trees.

The greatly increased demand for hardwoods since World War II has opened up vast new areas of forest to exploitation, with road builders following the logging crews. In more tropical countries than we care to count, a technique called clear-felling comes into play. Such practice is often only a part of various coordinated schemes, amounting to what the American conservationist Aldo Leopold has called skinning operations—a quantum leap beyond old-fashioned creaming. As for the clear-felling (or clear-cutting) the forest cover is completely removed. Scrubland remains, former undergrowth. Worse yet, timber is sometimes burned on the spot, as it is in ranching operations. Such approaches can totally consume the capital resource represented by the forest, with much waste of commercial timber.

A more recent variation of clear-felling, known as full-forest or any-tree-all-tree harvesting, resulted from the development of a technology in which up to one hundred or more species are chipped together to produce pulp for paper. This allows a forest to be exploited in a great hurry, with superb cabinet-grade hardwoods pulverized among the rest. Despite this appalling misuse of fine woods, full-forest harvesting has been heralded by some people as the possible salvation of the rain forests in that it concentrates total exploitation on a few areas, possibly in strips. Thus, as the usual rationale proceeds, mature forest will surround the bald areas and will spread out to regenerate the forest as before. If this regeneration did not occur, hardwoods could be replanted. We know of no such scheme that has worked as advertised.

Any controlled or managed regeneration of the forest after clearing, either in whole or in part, is the job of the forester—the person who most directly supervises the exploitation of the forest, an intermediate-level manager acting as the agent of the "authorities." But the forester must also give advice about proposed exploitation and its consequences, that includes saying "No!" He will also have to stand up and be counted in the conservation debate.

In an outspoken address to the World Forestry Congress in Jakarta in 1978, Jack Westoby, a leading FAO forestry expert, noted that foresters will be required to decide just where they stand: on the side of rapid exploitation; or with sustained production with all the conservation measures this entails.

The forester's pivotal role in global resource management will be made far more effective as honorable alliances are struck between business interests and the representatives of scientific conservation. These two groups share a profound interest in creating the highest levels of economic production consistent with the protection of nature itself.

Perspective must be restored if prosperity is ever to be achieved in many of the lands where threats to forest resources are most serious. In fact, positive change is needed if dozens of nations are not to sustain irretrievable damage to their resource bases. In 1977, K.F.S. King, FAO Assistant Director General in Charge of the Forestry Department said it very well:

The world cannot afford the clearing, burning and poisoning of trees practised by foresters in almost every tropical country fortunate enough to possess natural high forests.

Just how serious is the potential damage? One must state categorically that while the loss of a single species is cause for concern, the loss of a million species of plants and animals in the next 30 to 50 years would represent a biological disaster without parallel in the whole of evolutionary history. The distinguished botanist Peter Raven has estimated that each plant species which becomes extinct in the tropics takes with it as many as 30 dependent organisms. Because very few of these interdependent species have yet been studied, they may be lost before their potential value to mankind is ever known.

Unless we stem the tide of destruction that besets the world's rain forests, it is only this generation that will have the opportunity to study these little-known species. For our children it will simply be too late.

An Overall Assessment

As we have seen, the overall picture of exploitation of tropical rain forests is very patchy and varies from region to region. Cattle-ranching, for example, poses relatively little threat to the rain forests of Southeast Asia. Yet for parts of Latin America it is the major danger. Large-scale commercial logging has already cleared out many of the forests of tropical Africa and Southeast Asia, the two regions that have so far supplied most of the world's demand for tropical hardwoods. The impact of modern mechanized cutting is far less in Latin American rain forests, with logging there of a much more general character; although the signs point to an early change.

Tropical deforestation has been going on for a very long time and its acceleration, in most cases, coincided with the establishment of plantation economies. These are aimed at exporting crops to an ever-growing world market—sugar, coffee, rubber, palm oil, tea, tobacco, and others. Such agricultural experiments have been successful in the sense that they produce sustained yields on nutrient-rich soils. On the other hand, conversion of forest on unsuitable soils—usually the older soils poor in nutrients—can only lead to short-term gain and eventual disaster such as catastrophic erosion.

On a world scale, most of the forests on rich soils (such as those of the Mekong delta and Java) have already been converted for the cultivation of rice or other crops which support millions of people. We cannot reverse history, nor would we wish to do so.

Most of the world's remaining rain forests are found on poor soils and are subjected to extensive conversion. They are not generally suitable for large-scale clearing and agriculture. Just as agricultural practices of the developed world cannot be applied successfully to developing countries, so the exploitation policies that have been applied to the forests of the Old World are often unsuitable for the New World. It is not yet too late to learn this lesson.

There seems to be wide agreement amongst many foresters that one of the best hopes for reducing the rate of agricultural encroachment into tropical forests is to provide a viable alternative to forest settlement by improving agricultural productivity. Such an option would also help to preserve genetic resources and protect wildlife as well as the habitat of the 200 million people who live in and near the forests. As it is, agricultural productivity is below the capacity of the land in most of the developing world.

Louis Huguet, Director-General of France's Centre Forestier Tropical, boldly states that the main threat to the tropical forest is the primitive state of agriculture and grazing, rather than commercial, export-oriented logging. He calls for a new agricultural revolution such as that which occurred in the eighteenth and nineteenth centuries in Europe. He also draws attention to the very low level of forest management and the fact that for every ten acres of forest destroyed, at best only one acre is planted.

Yet reafforestation does have its success stories: Brazil with nearly ten million acres in forest plantations already supplies nearly sixty percent of the country's industrial wood needs while more than six hundred and ninty million acres of Amazon forest accounts for only ten percent of such use. Indeed, if it were not for the threats of agricultural encroachment and the policy of the government to open up the Amazon for development—most of it set on a course for failure, as we see it—it should be possible (at least in theory) to preserve the remaining Amazonian forest as a gigantic biotic reserve! Indeed, in view of the enormous economic value of the timber contained in the forest, it could be argued that this would be a wise investment in an appreciating capital resource. Estimated at early 1980s international prices for hardwoods, this figure is more than $1 trillion. The basic problem of the tropical forest remains, however, that of devising means of spending the interest while retaining the capital.

Duncan Poore, a recent Director of the Commonwealth Forestry Institute, wrote that we need a fundamental change of attitude from those with the power and the money. Currently, many such people view tropical deforestation as a positive step towards providing food, living room, and fuel for millions of poor people. At the same time, some of these people see the forest as a source of instant revenue, providing vital foreign exchange. Rather, the long-term profit lies in proper use of the forest as a capital reserve, with reafforestation as an investment for the future. In other words, the time is ripe for a change from exploitation to resource building. Sustainable use is the proper goal, rather than alternating phases of plunder and emergency remedial treatment. It will take a long time to build up again the forest resources and it will require a massive investment. Once they are planted, though, the trees will do most of the work.

Now we arrive at perhaps the most delicate and sensitive conservation message in the whole book. Tropical countries have sovereignty over their forest resources and clearly intend to use these resources to their own benefit, and as they see that benefit. They do not take kindly to being reminded by outsiders that much of their forested land is a world heritage, especially when that viewpoint is put forward most often by the rich nations' club—especially Western Europe and North America—most of whose agricultural development has been on previously forested land.

Yet all too often this sovereignty is ceded to those very outsiders for short-term gain.

Brian Johnson writes in a recent *IUCN* (International Union for Conservation of Nature and Natural Resources) *Bulletin* that while many of the countries which possess tropical forests are beginning to share concern at the *effects* of deforestation, they are slow at cooperating in action to tackle its *causes*.

We would all be well advised to cooperate in providing the scientific information necessary to allow national forest strategies to be developed—strategies that will involve all the users of the forest. We speak not of just the foresters, concerned as they are with growing or cropping tall straight trees, but of agriculturalists, economists, hydrologists, ecologists, planners, sociologists, and anthropologists. All of these people should be involved in the decision-making process. And a vast increase in the amount of international aid must be applied if we are to solve these problems.

Forests are for the people. Let us save them for the people.

Banana leaves serve as umbrellas for these Indonesians. Unable to afford many manufactured goods, rural people of the tropics often adapt plant parts to their daily needs.

Wetlands

THE HELMAND RIVER OF AFGHANISTAN, WITH no outlet to the sea, pours its waters into seasonal lakes in the interior. Opposite, an Afghan cuts reeds at one of the dried up lakes. With these and mud, he can repair and rebuild his family's homestead.

Little more than a thousand miles to the west, lies another remarkable Middle Eastern marsh. It provides a haven for a people living near the Shatt al Arab river. Though they are otherwise insulated from time, in the early 1980s sounds of a war between Iran and Iraq intruded into their domain. Iraqi citizens, these traditional people have learned to cope with the marsh on its own natural terms. The original settlers probably entered the marshes five thousand or more years ago. Their descendants' fascinating way of life has been graphically described in Wilfred Thesiger's classic of travel writing *The Marsh Arabs* and their history has been brought up to date in Gavin Young's *Return to the Marshes*.

This South Dakota farm pond feeds and shelters migratory swans. Glacial action spangled the prairies of Canada and the United States with thousands of such pocket habitats.

These thirty thousand Shiite Moslems live on artificial islands among six thousand square miles of lagoons and muddy banks. Their marshes lie near the junction of the Tigris and Euphrates, two almost legendary rivers which mingle and flow over flat countryside and form the Shatt al Arab waterway at the border of Iran and Iraq. The villagers, who have named themselves the "Ma'dan," weave mats from giant reeds, and fashion them into high-roofed dwellings. They also spear carp, cultivate rice, and herd water buffalo. They regard this beast—the source of milk, butter, and meat—as the symbol of their prosperity.

Trachoma, tuberculosis, dysentery, and the parasitic infection schistosomiasis, make the lives of these people far from utopian, and few doctors visit them. Yet the Marsh Arabs of southern Iraq have one advantage over the rest of us: the Ma'dan retain their Eden, watery though it is. They experience the herons, ducks, Siberian geese, cattle egrets, and other marsh waterfowl among bullrushes in the domain they have chosen.

Curiously, the marsh people's ancestors—living at the base of what has been called the Fertile Crescent, where agriculture emerged—may have "invented" civilization as we know it. Close at hand lie the sites of Sumer and Akkad, sources of cuneiform tablets that tell of the patriarchs who first stabilized sodden earth with mats of reeds from these same swamps.

This is one of many debts we owe wetlands but seldom realize, acknowledge, or repay. Ecologically, this is in error. And in anything less than a book-length treatment, the various types of wetland and the individual areas can only be sampled. It is in their nature to be scattered, and often small in relation to the vast life zones, in which and around which they most often occur. Wherever we live, even in deserts, as we have seen above, wetlands are seldom distant. Moreover, in nature's economy and by man's own standards, the living wetlands usually more than earn their keep. They provide breeding grounds for birds, land animals, and especially for fish—marine and freshwater species alike. In

fact we often see such animals and lose sight of their vital habitats.

Wetlands also profit us by filtering and recycling the chemical wastes that gush from factories and drain from cultivated land. The water carries unusually large amounts of nitrogen and phosphorus, chemicals taken up by the plants and used as nutrients. Certain wetlands can also function as denitrifying agents, removing specific chemical compounds from the water, just as certain muds rich in bacteria can apparently remove heavy metal pollutants such as chromium, cadmium, lead, and zinc from marsh waters. Some buoyant weeds, such as the notorious water hyacinth, even act as floating filters. Partially cleansed water seeps slowly into the aquifer, the subterranean source of ground water, where further natural filtration removes more impurities. In the face of huge quantities of chemical pollutants, however, the wetland's ability to cleanse itself does break down.

The familiar odor of rotten eggs is actually a

In the interior lowlands of Afghanistan, freshwater lakes support rich marshes. Reeds harvested by the farmer at right will go to repair his family's mud-and-wattle dwelling.

sign of health for the wetland ecosystem. It is produced by the metabolism of bacteria which flourish in waters poor in dissolved oxygen, a natural purifier. These organisms are able to transform toxic solids into a gaseous pollutant, hydrogen sulfide, which the winds sweep away. In Europe and North America, alder and sweet gale draw gaseous nitrogen from air or water and transform it into ammonia, a plant nutrient. The nitrogen thus "fixed" becomes a part of the ecosystem.

The massed greenery lessens the harmful effects of local flooding or helps prevent it altogether. After heavy rainfall, vegetation slows the flow of water, preventing its rapid run-off into neighboring areas. Released slowly into the streams of the marsh, the impounded water also has the opportunity to percolate into the aquifer and thereby add to the supply of ground water.

Wetlands often appear as an edge or a transition zone—scientifically termed an ecotone. Wet and dry mingle here, usually with one or the other in temporary ascendancy. And it is the flux between wet and dry, and the resulting shallow water filled with nutrients and dissolved oxygen, that helps give wetlands their ecological value—value that is too often ignored. These useful plant communities range from the salt marshes and mangrove swamps of the tidal zones to the mountain beavers' mires and from the Arctic's quaking muskeg to the flood plains and deltas of great rivers.

The Mangrove Swamps

Mangroves are trees and shrubs adapted for life in tropical and subtropical coastal areas. They belong to several botanical families, but all share a very uncommon ability to thrive with their roots in water—often brackish or even seawater. Their large masses of vegetation stabilize the soil, thus preventing coastal erosion.

Tropical mangrove swamps, or mangals, often serve as the border between the dry land and the sea, and are formed on estuarine mud that is regularly inundated with tidal water. Air temperature usually determines their geographical range, for they grow best where the temperature does not fall below 68° F. Species of the mangrove genus Avicennia tolerate 50° F in Brazil and New Zealand, and manage in temperatures as low as 54° F in Florida and at the northern end of the Red Sea.

Mangrove vegetation represents eleven families and sixteen genera, with forty-four of the world's fifty-five species occurring in the Indian Ocean and Western Pacific region. In the geographical sense, these mangal types are found in two great areas. The Indo-Pacific region includes the coastal and lagoon mangroves of East Africa, the Red Sea, areas of the Arabian Sea, India, Southeast Asia, southern Japan, the Philippines, Australia, New Zealand, and Pacific islands. The second major range extends from West Africa, across the Atlantic, and along the Pacific coast of tropical America down to only 40° S. latitude, and the Galápagos Islands.

The mangroves themselves grow into trees except at the northern and southern extremes of mangal, roughly Louisiana and New Zealand, where they appear as bushes. The intertwined prop, or aerial, roots help characterize the forest. In *Rhizophora*, woody breathing organs or pneumatophores rise from the ground, and *Bruguiera* has distinctive "knee roots." *Rhizophora* and *Bruguiera* produce large, hanging seedlings while those of *Avicennia* resemble beans. Mangal areas contain few lianas, unlike some nearby jungles in which these trailing vines link tree to tree. There are also few epiphytes, or air plants, though members of the coffee family (Rubiaceae) grow in the canopies of some mangroves in the Philippines and Indonesia.

Biological activity in the mangrove swamp reflects the ebb and flow of the tides and the varying influences of coastal currents. Plants as well as animals are caught in the flux. Moving water carries particles of silt, sand, peat, and even coral from reefs, all of which eventually build a seed bed and anchor for roots. The lower trunks of mangroves are periodically immersed. Because they absorb the force of wind and tidal surge, mangroves protect coastlines. Their dense root mats counteract soil erosion. The trees themselves bend but seldom break during tropical storms. As Hawaii possessed no native mangal, the red mangrove (*Rhizophora mangle*) was introduced in 1905 to stabilize the shoreline on Molokai.

Year in and year out, mangroves produce considerable organic matter. These swamp forests probably transform the sun's energy into as much organic matter per acre as good farmland or the evergreen coniferous forests of the Northern Hemisphere, and they support a variety of wildlife. The landward or interior portion is home for frogs and crocodiles, whereas the ocean edge affords muddy footing for shrimp, crabs, and molluscs. Prop roots are colonized by

oysters and sponges. A network of minor waterways, that may drain as the tides change, allows worms, crustaceans, and fish to move about. Overhead, birds search for food and shelter.

The Caroni Swamp of Trinidad is one of the Caribbean's renowned mangrove areas, due mostly to scarlet ibis (*Eudocimus ruber*) which nests here. Situated on fourteen thousand acres just south of Port-of-Spain, on the deltas of the Caroni and Blue Rivers, it supports ten thousand scarlet ibis in the Caroni Wildlife Sanctuary, with herons, egrets, and other waders also breeding here. At periods of low tide, the ibis consume various crustaceans and molluscs.

The most extensive mangal community in the world is located at the Indian subcontinent, fringing the delta of the combined Ganges, Brahmaputra, and Meghna rivers. Its most abundant growth occurs in an area collectively known as the Sundarbans, covering twenty-three hundred square miles. The region holds 85 percent of India's mangroves, supporting nearly twenty species and, amongst the animal species, the royal Bengal tiger finds refuge here. According to one estimate, half the original forest has been lost during the past two or three hundred years.

Across the subcontinent from this delta, local exploitation for firewood and tannin, and the consumption of seedlings by cows, has endangered the west coast's stands of *Rhizophora* timber. Mangrove species are, in fact, of considerable economic value. Their prime use is as a firewood of high caloric value, prized by villagers who often come by canoe to cut it. *Rhizophora mucronata* and *R. apiculata* provide timber twice as durable as teak for coal-pit props in Nigeria.

The strong, dense wood of *Rhizophora mangle* is resistant to termites and is made into wharf pilings, planks, and poles, though shipworms can weaken it. *Bruguiera gymnorrhiza* is similarly used for poles in India and planks in East Africa. Local people take advantage of the high tannin content of several mangrove species; this organic acid is used for tanning leather.

In Malaysia, the fruit of *Bruguiera caryophylloides* soothes sore eyes. The sap of *Avicennia* species is employed for contraception while the seeds provide an ointment. The leaves of *Acanthus ebracteatus*, a denizen of Malaysia's mangal but not itself a mangrove, are used by local people to treat rheumatism and its roots applied to alleviate poisoning from snakebite and the action of other toxins.

In areas of transition from salty to brackish water, the fern *Acrostichum* abounds, and in the Indian Ocean and Western Pacific region, the *Nypa* palm will produce extremely dense colonies that provide materials for shelter, food, and protection from erosion.

Salt Marshes

These salt-tolerant plant communities occupy coasts of most continents, and thrive in the area between the low and high tidewater levels in the temperate zones. From a distance they sometimes resemble great lawns. Maritime or coastal salt marshes have been differentiated around the world according to the dominant species of plants.

Arctic salt marshes of North America, Greenland, Europe, and the Soviet Union are predominantly covered with the grass *Puccinellia phryganodes* and some sedges of the genus *Carex*. Northern European marshes occur from Spain and Portugal north to the English Channel and across the North Sea coastal areas of the Low Countries and the more brackish Baltic Sea. In general, the dominant plants of northern European marshes are annual species of *Salicornia*, the *Puccinellia maritima* grass, and the rush *Juncus gerardi*. The Channel marshes are notable for *Spartina townsendii*, a natural hybrid.

Marshes of the western Atlantic support salt marsh cordgrass (*Spartina alterniflora*) in tidal areas, and in areas slightly above the tides one finds *Spartina patens*, the salt meadow cordgrass. Other species colonize shallows of the Mediterranean and Caspian Seas. Large sections of salt marshes are also found on the Pacific Coast of North America, Baja California, Japan, Australia, and South America. *Spartina* grass in the marshes of eastern South America has evolved into two separable species, *S. brasiliensis* and *S. montevidensis*.

These marshes are generally unable to compete successfully in tropical areas of mangrove growth, yet the two nearly coincide at India's northwest coast, close to that area's previously mentioned mangrove community. Near the Kathiawar Peninsula lies the district of Kutch—a scorched and arid land bounded by the Gulf of Kutch to the south and separated from Pakistan and the Indian mainland by the Rann of Kutch. The Rann, a salt marsh the size of Massachusetts, alternates between wet and dry. Floods convert Kutch into an island during the rainy part of the year, but a sandy and salty desert emerges between monsoons.

The Little Rann of Kutch occupies the northwestern portion of the State of Gujarat, and comprises one thousand square miles of flat salt "wasteland," dry from November to June. Monsoon rains come from July to September, submerging the flats under a foot or more of water, with only occasional islands of vegetation. This remarkable habitat comprises the last major stronghold of *Equus hemionus khur*, an endangered wild ass.

Together, Indian marsh and mangrove support a biological cornucopia. Primary biological productivity is further enhanced by a coral community in the shallow waters. Reefs occur at the Kathiawar Peninsula, where coral heads are separated by patches of sand, mudflats, and mangrove. Intensive harvest of mangrove stems for fuelwood and leaves for fodder appears to have upset an ecological balance of long standing.

In France, more than four thousand miles northwestward from The Rann of Kutch, another famous salt marsh is found. A cooler cli-

Plants and animals of the Chinese marshland have been fitted into the overall design of this ceramic medallion mounted in a wall of the Imperial Palace in Beijing.

mate rules out the presence of mangrove in the Camargue where the Rhone River of sunny Provence loses itself in a sandy delta. At its heart lies a plain of nearly three hundred square miles, which attracts an unusual diversity of bird life. Nearly three hundred avian species congregate in the Zoological and Botanical Reserve of the Camargue, a wilderness of twenty-six thousand acres administered by The French National Society for the Protection of Nature.

Each species of bird chooses its habitat based on salinity of the stream or marsh. Brackish lakes spangle the Camargue, and to their islands come the gulls, terns, and plovers. Areas of brushwood along the river support egrets and heron, while sandpipers and snipe cluster around sweeter water. Flamingos, the largest European colony of these wading birds, favor brackish ponds and thousands of North African flamingos also find their place within this highly structured habitat.

Yet there is trouble. Each year, people of the productive Camargue rear prize bulls, produce a hundred thousand tons of rice, and extract more than 1.5 million tons of salt. The herbicides and hormones applied by the rice cultivators to their flooded fields work their way into the marshes. Water salinity fluctuates with the repeated flooding and draining of the ricefields. Worse yet, noise pollution—so repugnant to persons living beneath the flight paths of jet aircraft— similarly unnerves the flamingos at their Camargue nesting sites. Thousands of tourists on the beaches contribute their refuse to the rubbish heaps that mar this enchanting place.

The same threats—tourism, urbanization, and alteration of the water table through extensive drainage—have also threatened the salt marshes of southwestern Spain. Near the mouth of the Guadalquivir River, at the Gulf of Cadiz, a natural refuge holds two hundred species of birds, including flamingos, herons, avocets, rare white-headed ducks, stilts, storks, ibis, vultures, and 20 or so species of mammals, including wild cat, Spanish lynx, genet, deer, boar, and about 30 species of amphibians and reptiles.

The plight of the Spanish Imperial eagle exemplifies the problems now being faced by salt marsh denizens. Accustomed to nesting at the borders of the marshes, birds encounter noise and pollution, both by-products of the tourist industry situated near the Coto de Doñana wildlife reserve. Today, many of the eggs of the great eagle are sterile and the future of the species is in grave doubt. This European paradise, in which half of the bird species of the Continent have been seen, is described in Guy Mountfort's *Portrait of a Wilderness*. As a consequence of attempts to save the Coto de Doñana the World Wildlife Fund was established in 1961.

Further to the west, across the Atlantic, large salt marshes occur on the fringes of North America. Estimates vary, but salt marshes here may cover thirteen thousand square miles and are probably disappearing at about one percent every two years. About half of them occur along the Gulf of Mexico, on the edges of the Mississippi River delta in Louisiana. If present trends persist, a million wetland acres may be converted to other uses or just simply disappear as integrated plant communities during the closing decades of the twentieth century. The most serious threats to these salt marshes are sewage pollution, channel dredging and development, fill, chemical wastes from industry and farm, ditching and draining of wetlands, agricultural wastes, river impoundment, and flow control.

Tidal flow transports sediments needed for plant growth, and flushes away the metabolic wastes of plant and animal. As the tides fall and soil is bared to the atmosphere, the oxygen supply to organisms in the mud and sand increases. The cycle occurs once every 24 hours and 50 minutes on the coast of the Gulf of Mexico and twice during this period at the Atlantic and Pacific coasts. The animals are adjusted to this rhythm, and may seek food at high tide (oysters) or low tide (crabs).

Marsh vegetation fills the water with that biologically important though little celebrated substance, detritus: a mixture of decomposed plant materials and bacteria. Crustaceans consume it and their waste products are acted upon by the bacteria. Detritus is as efficient a fodder for shrimp as is the floating minutia grazed upon by the filter-feeders and the larvae and hatchlings of various creatures.

We have gradually learned the economic and ecological importance of salt marshes and other features of the continental fringes. Associated with estuaries, for example, wetlands can provide nursery grounds for commercially profitable fish and shrimp. For instance, the brown shrimp spawns in the Atlantic, out beyond the sandy barrier islands of the eastern United States. Juvenile shrimp move from the ocean to protected marsh areas between the islands and the mainland. As adults, they return to the sea.

Freshwater Wetlands

Swamps often become established in wet depressions that receive water from ground percolation, surface springs, streams, rain, run-off water, and other sources. Bogs, though, often result from glacial action—and are frequently found in Canada and the northeast and the upper Middle West of the United States. The arctic and sub-arctic regions of Eastern and Western Hemispheres abound with them. Generally there is extremely little flow of water in or out and few opportunities for water to drain.

The accumulation of disintegrated plants— detritus—fills many depressions. The weight of

Global demand for specimens of the carnivorous pitcher plant (Cephalotus follicularis) *has decimated populations in its native Australian swamps; it is now listed on Appendix II of the Convention on International Trade in Endangered Species of Wild Fauna and Flora (CITES).*

up to forty feet of organic material provides the pressure to form peat, with the world's peatlands estimated at as much as eighty-eight thousand square miles, an area nearly the size of the United Kingdom. Springy sphagnum moss grows on the surface of many bogs. Insectivorous plants such as sundews and pitcher plants often abound. They have evolved a curious carnivorous habit to enrich their supply of nitrogen, a plant food scarce in bogs.

The German-Polish Lowlands and Russia's Pripet Marshes were glaciated along with Norway, Sweden, and Finland. Lakes and marshes are common here. The Western Siberian Lowlands hold the Soviet Union's extensive Vasyugan swamp to the north and east of Omsk. It comes as a surprise that this region was not directly affected by the continental glaciation.

When the huge ice cap of the Northern Hemisphere melted a dozen millennia ago, the level of the sea rose and flooded land lying inland of the ancient beaches. New wetlands were thus born. In regions of North America not scoured by glacial action, such as the mid-Atlantic and southern states, comparatively flat land and shallow water support great coastal marshes, some of which are threatened.

The tropics also stand to lose several types of freshwater wetlands. One kind is drowned by floods, then becomes higher and drier as swollen waters ebb. Papua New Guinea's Fly and Sepik River swamps, are examples. Large-scale projects to drain parts of the Sudd, in the Sudan, have begun. Papyrus (*Cyperus papyrus*) dominates the vegetation of this huge swamp system which has a permanent area of thirty-eight thousand square miles and in the wet seasons swells, at the will of the Nile, up to seven times its base size. Overall, the area of humid tropical swamps is estimated at one hundred and thirty thousand square miles, an area roughly the size of Italy.

In South America, Brazil's Pantanal seasonally loses a quarter of its area. Conservationists are concerned that parts may be artificially drained. Although a new Pantanal park has been created, poachers trespass to take crocodiles, jaguars, capybaras, and otters. Other regions are also vulnerable. These include the area of varzea vegetation along the lower Amazon River, the delta and backwaters of the lower Orinoco, the upper Xingu River, swamps in the Território Amapá of Brazil, and wetlands of the Beni Department of Bolivia.

Clearly influenced by the climate of the tropics, and rightly extolled for the diversity of its plant and animal species and its great size, Florida's Everglades represent the most awesome freshwater swamp system in the United States. Several types of habitat are woven together within the watery scene of sawgrass and bald cypress trees.

The Everglades ecosystem as a whole encompasses thirteen thousand square miles, an area larger than the Netherlands. It can be said to originate at Turkey Lake, due west of Orlando, from which water flows by rivers and canals through several other lakes, and ends with the Kissimmee River feeding into Lake Okeechobee. In this watery corridor one encounters the Fakahatchee Strand, a splendid botanical experience and a place that should never be made more accessible.

A compass is vital, as the wader—variously from knee to thigh depth—may never return without such guidance. It is worth the risk. An almost unbelievable display of tropical plant life is eventually stumbled upon in the "Orchid Cathedral" of the Fakahatchee. In their bewildering myriads, the epiphytic orchids, ferns, and bromeliads hang on tree branches, and provide a serene, otherworldly aura. Over-arching limbs and mist arising from the water surface catch green-gold sunlight and create a shrine to the splendor of life upon our planet.

Ferns in the Cathedral include *Osmunda regalis, Asplenium serratum, Asplenium auritum,* and *Campyloneuron angustifolium.* The Strand is the only known location of the *Campyloneuron* in Florida. The club moss *Lycopodium dichotomum* occurs in the Fakahatchee, its only site in the United States, with a population of only two individuals.

Two species of the tropical pepper family, *Peperomia humilis* and *P. obtusifolia* appear. Festooned orchids include *Epidendrum rigidum* with its zig-zag inflorescence, *E. anceps, E. nocturnum,* species of *Pleurothallis, Ionopsis utricularioides, Encyclia cochleata,* and a species of the ghost orchid-*Polyrrhiza.* The primitive whisk-fern (*Psilotum nudum*) clings to a few branches. It is curious that *Tillandsia usneoides,* Spanish moss, is uncommon in the Fakahatchee. Other bromeliads, or air plants, are in evidence throughout, and amply compensate the spectator. The fuzzy-wuzzy air plant (*Tillandsia pruinosa*) makes a rare appearance.

The environment of the Everglades has been damaged over a period of decades in attempts to make the region economically productive and to enhance its suitability for human occupation. The most notable attempts involve drainage of areas in order to make them drier for agriculture. Efforts by conservationists have been aimed at holding the line: at stopping the spread of farms westward from Miami into the Everglades as Florida's population seeks to expand into new domains.

A better balance between economic and ecological functions of the wetlands can benefit millions of people. For instance, the Chesapeake Bay of the United States, one of the great estuarine treasures of the world, now produces at much less than its estimated sustainable capacity, and pollution contributes to the problem. Ongoing work at the Chesapeake Bay Center for Environmental Studies, part of the Smithsonian Institution, provides support for a recent initiative of the Federal Government to bring the Chesapeake into its own again.

Here as elsewhere, motivation for more effective conservation may come from the awareness that wetlands have special importance in assuring adequate supplies of clean water. Wetlands also provide food and fuel, and help to prevent floods and erosion. The need for positive change is greatest in densely populated areas, notably around large cities located near bays and estuaries, and in the crowded deltalands of the developing world.

The 15-inch flowers of the royal water lily (Victoria amazonica) *grace quiet pools in Venezuela, Brazil, and Paraguay. Buds and enormous leaves appear in the background.*

Green Oceans

PHOTOSYNTHESIS IS CARRIED out not only in land plants but in the upper layers of the oceans and seas. Myriads of microscopic phytoplankton as well as kelp and other seaweeds are involved. They exist in such numbers that they probably use more carbon dioxide than the land-based flora, perhaps 80 billion tons a year compared to half that figure for all terrestrial plants. This active absorption reduces the air's carbon dioxide gas, produced naturally by plants and animals throughout their lives. Nature's balance can, however, be thrown off. A major offender may be all the coal, oil, and gas that we burn to provide for our wants, since industrial and domestic combustion liberates carbon dioxide gas into the atmosphere. Concentrations also increase when tropical forests are burned to clear the soil for agriculture.

Whatever the source of this invisible gas, the total amount of carbon dioxide in the air influences the naturally occuring "greenhouse effect." Air at the surface of the land and the sea is warmed at the same time that clouds and other atmospheric phenomena insulate our world from the coldness of space. The higher

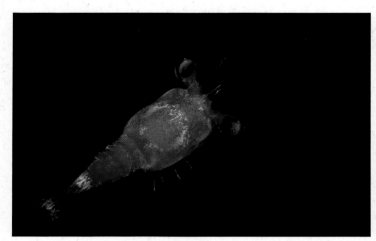

Rivers reach the salty Gulf of Mexico by way of passes between barrier islands, as at Corpus Christi, Texas, right. Beyond the Gulf is the Gulf Stream where the likes of the decapod shrimp, above, feed on microscopic algae.

128

the concentrations of carbon dioxide, the more that heat from sunlight and other sources is trapped. According to theory, it is also conceivable that an ever-thickening blanket of cloud could envelop the planet. Temperatures would continue to climb and the level of the oceans might rise as polar ice melts. Some trends seem to indicate that this may be happening, and scientists continue to seek confirmation or denial of their suspicions.

Since carbon dioxide nourishes marine algae, any abrupt alteration of saltwater plant populations could provide early warning of climate change. Fortunately, multisensor satellites can identify and measure chlorophyll in the oceans. The new method captures immense amounts of biological data over vast expanses of oceans and also from places where rivers flow into the salty sea.

Below; shallow waters of the Great Barrier Reef of Australia support such creatures as Lysmata grabbami, *a cleaner shrimp. Continental shelf waters are often richest in both plant and animal life.*

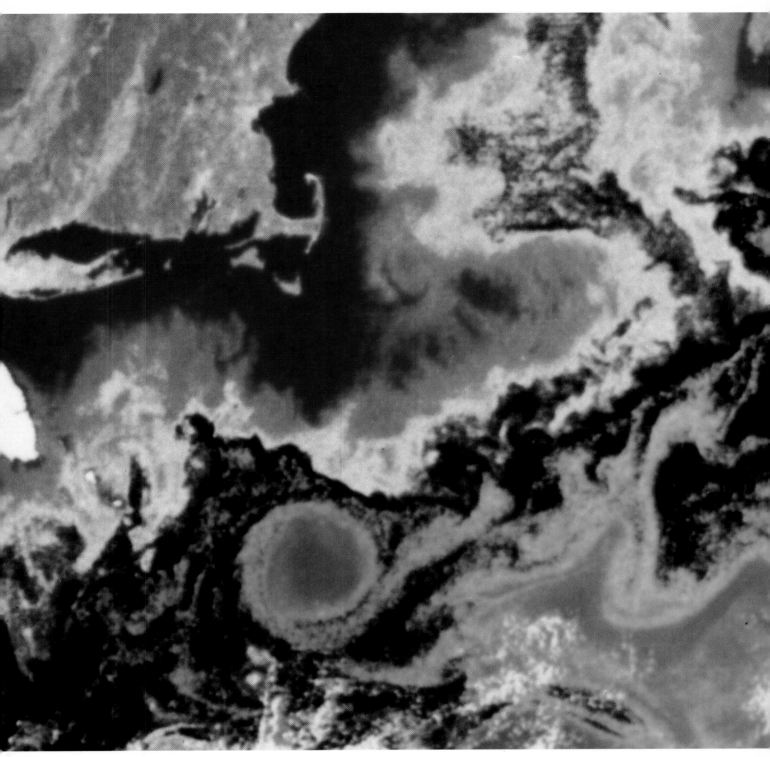

*Above; satellite view reveals levels of
chlorophyll in Atlantic waters from
Long Island to Cape Cod. Reds and
browns are richest, with the blue
Gulf Stream and its detached eddy
poorest in plant life.*

Fair Isles

In 1769, during Captain James Cook's first Pacific expedition, artist Sydney Parkinson painted the ornamental legume Sesbania tomentosa, opposite, from the Society Islands (now in French Polynesia). The W. Hodges painting below shows ships Resolution and Adventure from Cook's second voyage, off the coast of Tahiti.

ISLAND LIFE HAS SPECIAL CHARACTERISTICS that have attracted much attention from scientists. The pattern of variation in closely related finches and other animals in the Galápagos helped Charles Darwin frame his ideas of evolution and natural selection.

The scientific fascination with islands began nearly a century earlier when Sir Joseph Banks and his retinue of naturalists and botanical draftsmen accompanied Capt. James Cook on his historic circumnavigation of the globe. Banks was a wealthy 25-year-old when "Endeavour" set out in 1768. In later years he was largely responsible for the scientific foundation of the Royal Botanic Gardens, Kew.

While his main interest was in botanical specimens, Banks and the "scientific gentlemen" on board Cook's ship collected a wide range of "natural historical" materials, wrote their journals, painted landscapes, plants, animals, and minerals, and keenly observed as they visited island after island. Their southernmost landfall was that largest of all islands and smallest of all continents, Australia. From their era onward, research into the peculiarities and mysteries of islands and their flora and fauna has gathered momentum.

What then is so special about islands? We know that early adventurers were struck by the fact that semi-tropical and tropical islands held so many more plant species, area for area, than the continents. They were also keenly aware of the commercial opportunities, and those early explorations were meant as much to enhance and enrich empire as to expand it. Ships often returned home as floating treasure chests, with perhaps the most enduring wealth reaching Europe in the form of seeds, roots, and live plants. This massive transfer of genetic material between the Old World and the New laid the foundations for much of the planet's agriculture.

Why should this extraordinary diversity of species be found on islands? One of the great island researchers, Sherwin Carlquist, identifies two main reasons for the biological richness. The first deals with dispersal, and the second with evolution. He writes, "Ocean islands are, to me, archives of the fact of long-distance dispersal and the way in which it works and has worked."

This far scattering of seed by bird, air, and ocean has fundamental consequences when taken into account with a second concept, also Darwin's, that of adaptive radiation to island situations. Once a newcomer (either bird, beast, or plant) becomes established and reproduces on an island, evolution of new variants can often be explosive. Carlquist cites an example from Hawaii. On Oahu alone there are nearly 130 species of *Cyrtandra* (relatives of African violets and gloxinias). They may have arisen from only one or possibly, several individuals.

For its livelihood, each new adaptive form— potentially a new species—is related to some narrow aspect, or niche, of the island's ecology. Each species tends to be locked onto its own island, just as surely as the historical Robinson Crusoe, Alexander Selkirk, was confined to the islands of Juan Fernández. Competitive ability suffers in such evolutionary backwaters. Some of the newly evolved species would die out beyond their home ground. Without special protection, local plants have often been driven to extinction before the invasion of aggressive, alien species.

Islands are very fragile paradises indeed. By way of illustration, we can look at the early movement across oceans and seas as people searched for new lands. Before Europe's Age of Discovery, such island hopping continued at a steady pace. Seeds, plants, and animals were taken for food on the voyage and for breeding ashore. Such pre-Columbian migrations seldom if ever brought the scale of destruction that we have witnessed during the past 200 years.

In the era of European exploration and colonization, seafaring men often released goats and other animals to allow them to breed in the open and thus provide sustenance for seamen who might visit or be shipwrecked in the future. This was a sensible and even correct policy in centuries past. It went wrong through the lack of an overall plan and control.

St. Helena, a small island in the South Atlantic and place of Napoleon's final exile, provides an example of how quickly man's urge to travel and to take along his familiar food can bring about dramatic consequences. Goats were introduced here in 1513 by the Portuguese. In 1875, the great Victorian botanist Joseph Dalton Hooker estimated that there must have been more than a hundred endemic plant species— ones known no place else on earth—a "wonderfully curious little flora." Only 40 endemics remain on St. Helena.

Sadly, many other islands suffered the same fate as St. Helena. One was Philip Island, a tiny speck in the South Pacific which was visited, along with nearby Norfolk Island, by Capt. Cook in 1774. Botanists on the expedition collected many new species there, including *Streblorrhiza speciosa*, the Philip Island Glory Pea. It gave horticulture a beautiful and exciting new greenhouse plant. Today the plant no longer exists, either on the island or in cultivation—it has gone forever!

The Glory Pea disappeared because Philip Island was altered beyond recognition in the wake of human enterprise.

Today the island is essentially a colourful desert . . .

as authors of the booklet "The Conservation of Norfolk Island" indicate. Here, the physical erosion of the island's soil has coincided with the genetic erosion of the island's species. With

133

Islands of various South Pacific chains have fared quite differently under the hand of mankind. Right, the exquisite Rose Atoll lies surrounded by a wide pink coral reef at the eastern end of American Samoa. Designated as a National Wildlife Refuge, the atoll is inspected periodically to ensure the continued inviolability of its flora and fauna. Below, a fairy tern creates an ethereal presence in a Pisonia grove which was spared when Takapoto Atoll, one of 80 islands in the Tuamotu Archipelago, was turned into a coconut plantation. Opposite, vegetation on the spectacularly steep-cliffed Fatuhiva Island, in the equatorial Marquesas Islands, has been severely damaged by wild sheep, goats, donkeys, horses, and cattle. Today bracken and grasses dominate the knife-edged ridges.

the world that are in danger of vanishing forever as a result of threats to their habitat.

Whether done by accident or design, and whether on islands large or small, the introduction of domestic animals has brought particularly disastrous consequences. For instance, accidental introductions of rats and other animal "weeds" have destroyed for the most part the flightless birds and many other native animals of the world's islands. Alien plant species have been introduced in a similar fashion. Many weeds accompanied immigrants along with their crop seeds and fodder for the animals. Species from the temperate zones can grow and reproduce rapidly in hot country, and their seeds are often the first to germinate on broken soil.

New Zealand was a ripe target for alien plants. The British species that the settlers brought spread with vigor, invading so many ecological niches that today it is often difficult to find a single native species. The landscape is still green, but its flora is less diverse. Today, broom (*Cytisus scoparius*) and gorse (*Ulex europaeus*) cover many hills. No doubt these species, though alien to the new land, helped make the settlers feel at home. Yet the loss of native species is dramatic and widespread.

New Zealand is also an example of how other human influences have changed the landscape. Forest probably covered up to 80 percent of the land before the arrival of the Polynesians in about A.D. 1350. As David Given points out in his book, *Rare and Endangered Plants of New Zealand*, this figure had been reduced to 55 percent by the early 1800s, when European settlement began to spread. Now only 23 percent of the original area remains wooded.

The removal of trees helped to create some of the most productive sheep farming land in the world, a major contribution towards meeting the universal demand for protein, but early production was undeniably won at the expense of the native flora. Agricultural improvements can increase the yield of today's farm acreage,

enough energy and determination, a small green habitat like that of Philip Island could be reconstructed. But once extinct, a species can never arise again.

Covered in scrub, with dense forest in the valleys, this island provided an obvious "natural farm" for goats, pigs and rabbits. They were introduced to feed a penal settlement set up by the British some years after Cook's visit. Today, rooted in pockets of what must pass for soil on Philip Island, a few forlorn trees endure. Clinging to the yellow-and-red ground, *Hibiscus insularis* still hangs on and we hope that it too can be prevented from becoming extinct. This is just one example of the thousands of species across

and must do so, if we are not to destroy the last remnants of the wild landscape. Highly productive agriculture is now possible and increasingly important to the well-being of all countries. And for the sake of conservation, such economic advancement should encompass the development of parks and reserves.

Thankfully, New Zealand has set up reserves for such majestic trees as the kauri (*Agathis australis*), which is almost extinct today. Those areas not totally cleared for agriculture were, in the main, selectively logged. Some species adapted to the *Agathis* forests were almost driven to extinction as well. As was true with the actions taken on St. Helena and other islands, each step seemed right at the time. At least today, the folly of going too far without sufficient planning has been recognized and corrective ac-

Hawaiian hills have been topped to create irrigated pineapple fields. Native plants have been threatened as every suitable acre has been converted for cultivation of the delicious bromeliad, member of a botanical family of tropical and subtropical America. Dendrobium secundum, *an orchid of Indonesia and the Philippines, bears a waxy flower in the shape of a tube. Like many orchids it is an epiphyte, or air plant, not requiring soil.*

Death by overgrazing on Mohotane Island, in the Marquesas: the remains of a feral sheep, starved after eating every available leaf.

means by which New Zealanders can survive and flourish. Sustainable development is essential. Sustainable development will be achieved when conservation is fully integrated with development and the two are no longer viewed as mutually exclusive or opposite ends of the spectrum. Development based on conservation objectives uses resources in a sustainable manner, ensuring the long-term viability and growth of New Zealand society.

One hopes that New Zealand's initiative may serve as inspiration for other island systems. It certainly fits in with the World Conservation Strategy and thus has much to recommend it to mainlanders as well.

Another Polynesian outpost, the Hawaiian chain, closely fulfills many an idealized view of paradise. Yet Hawaii also presents a disturbing picture of a very rich flora under great stress, even today, when appropriate plans could and should be developed to secure the natural wealth of all the islands. Ayensu and DeFilipps' 1978 *Endangered and Threatened Plants of the United States*, lists more than eleven hundred Hawaiian species, subspecies and varieties as Endangered, Threatened, or Recently Extinct. This represents no less than half of Hawaii's twenty-two hundred species, subspecies, and varieties. Even more dramatic is the fact that 646 species have been identified as Endangered— more than 29 percent of the whole flora!

It is to be hoped that a comprehensive conservation strategy will be developed and actively implemented before this island paradise becomes famous as an example of conservation failure through lack of caring and action.

Off the coast of Africa, the Canary Islands are another tourist attraction with a history of ravaged vegetation. After centuries of exploitation, little remains of the celebrated laurel forest. The first deliberate large-scale deforestation began on the island of Tenerife when sugar cane was introduced as a crop early in the sixteenth century, after first being introduced in neighboring Madeira in 1419. Space was needed to grow it and large tracts of land were cleared: the

tion taken. This could have happened earlier! The stimulus provided by these great trees will continue throughout New Zealand's future as an inspiration even while remaining a symbol of this nation's development.

The first paragraph of the first chapter of the 1981 proposal for a New Zealand Conservation Strategy called "Integrating Conservation and Development" deals with a proper balance between resource use and conservation:

Human aspirations and survival are closely bound to the land and all its resources. Careful development of those resources can lead to economic growth, improved living standards, employment opportunities and increased well-being in the broadest sense. Depletion, destruction and over-exploitation undermine the very

downed timber was used to fuel the refining. The natural vegetation of these islands was further affected, often totally destroyed, by the cultivation of tomatoes and grapevines, the terracing of the slopes for these and other vegetable crops, the widespread cultivation of bananas in coastal areas, and the introduction and spread of many Mediterranean weeds.

This all led to a lowering of the water table, and increased aridity. Massive urbanization and tourist developments have added further scars to this already heavily defaced landscape. Although it is too late to reverse or even repair most of the damage, an enlightened island government runs and finances one of the world's finest conservation oriented botanic gardens, the Jardin Botánico Canario Viera y Clavijo, on Gran Canaria. It houses magnificent living collections of Macaronesian plants. Alas, very few of the tourists who come by the planeload from central and north Europe to enjoy the sunshine bother to visit the garden and sample some of the riches of the plant life on these islands, over a quarter of whose species are endemic.

Such instances highlight the kinds of problems that the plant world encounters everywhere and suggest possible remedies. They also present areas where scientific knowledge and financial resources are available.

Action plans can be readily undertaken in the developed nations. Getting started will be much more difficult in other similarly threatened areas. We refer especially to such important members of the international community as Indonesia and the Philippines. It seems unfair that tropical soils are so vulnerable to damage, but it is a fact of life. Neither is it a myth that islands can become overpopulated and thereby almost literally eat and burn up their natural resources.

In tropical islands, we believe, development plans should be built up in such a way that an expanding human population can be sustained by the wise use of natural resources. Progress in agriculture and forestry must provide for a sustainable yield—this point is vital. So let us urge that sufficient money be made available to launch suitable and achievable plans. All too often this does not happen.

There is another important aspect to consider. The World Conservation Strategy, covered in the last chapter before the Epilogue, identifies islands as major areas of conservation importance in "the programme for genetic resource

areas." One of the largest islands in the world, home of the Malagasy Republic, is a prime example—a relatively poor country with a great pride in its unique flora and fauna.

This wildlife wealth developed after the Malagasy Republic's island, Madagascar, broke away from the African mainland about 100 million years ago. This ancient separation allowed the evolution of a highly distinctive flora and fauna. The island holds twelve thousand plant species, five times more than Hawaii. The whole continent of Europe possesses only a few more.

Moreover, 80 percent of the plant species are endemic to the island, a phenomenal figure. All this botanical treasure fits in a space the size of France or of California with the adjoining Mexican State of Baja California. Unfortunately, 80 percent of the original vegetation has been destroyed, so these twelve thousand species of plants most likely represent only a fraction of the vast original flora.

A world of bizarre plants, Madagascar today ranges from dry desert to moist tropical forest. It also possesses a rapidly growing human population in need of the essentials, and more, for life.

In *A World Like Our Own*, Alison Jolly sums up the problems admirably:

> It is . . . a question of balance. If there are the right number of cattle, they find rich hay in the western savannas and feel no need to invade the forest. In the dry season they live on the "green bite" (the new grass shoots that spring up when fires pass). Some fire, and some green shoots are clearly necessary. . . . More people want more cattle and more cash. Cash crops in this region mean shifting cultivation of peanuts and therefore cutting the forest for fields, or else irrigated cotton on black alluvial soil in the stream beds. In turn, the cotton fields involve keeping the cattle out of what was once rich pasture, forcing the displaced herds to put still greater pressure on the dry grass and woodland. . . . Within the forest the cattle take shoots of favoured tree species, and, in time, the whole woodland becomes sparser, drier, sunnier. It is then more susceptible to fire, which can roll back the boundary tens or hundreds of metres in a year.

Later in her book, Alison Jolly shows how forest fires are destroying this unique country's

long-term chances. In a horrifying paragraph, she describes a principal threat:

One quarter of Madagascar burns each year. Fires crackle across the grasslands, driven by the prairie wind. Fires necklace the hillsides, chains of scarlet jewels in the night. Fires invade the plantations of pine and eucalyptus, destroying years of foresters' toil. The fires are set deliberately, then burn uncontrollably, stopping only when the wind drops or when a cliff face or patch of bare soil checks their course. After the fires more green shoots poke upwards from the charred tussocks of the grass—survival for the zebu, which are the peasants' pride.

The number of zebu cattle on the island is estimated at 10 million—greater than the human population—and they play a major role in the life and rituals of the people. They also contribute heavily to the loss of plant resources, and particularly of the primary plant resource—

ricultural lands are essential to top-up that feeding and life-supporting capacity. This still leaves many areas for "wild" plants and animals to live—where it would be economically insane for man to further disturb these precariously balanced and highly vulnerable ecosystems. As we suggest throughout this volume, wild places can be seen as biological savings banks, places of great value for the future.

In those regions where intensive or marginal agriculture is carried out, there can still be designated wild reserves to protect the quintessential botanical wealth—the genetic diversity represented by native species. It is not something that would merely be nice to do. It is essential for the long term, not only for the relatively new Malagasy Republic but for us all.

It is pleasing to be able to report that discussions have been held between the government of the Malagasy Republic and a team from the International Union for Conservation of Nature and Natural Resources (IUCN) and the World Wildlife Fund (WWF) to draw up the framework of a national conservation strategy.

The Malagasy Republic's environmental dilemmas transcend national boundaries, as do the ecological issues of other island systems. Money is needed to aid the Malagasy people to develop their own conservation and development initiatives. Furthermore, their wondrous island in the Indian Ocean is only one of hundreds of islands and many governments that face economic problems associated with their natural heritage.

What is needed is a vision held in common by all mankind, one that helps us break free of the we-and-they syndrome. Though each of us is primarily responsible for his or her own home ground, in a real sense this world belongs to every one of its inhabitants. Thus each person has some national, as well as some international, responsibility. The concept is not so complicated. Besides we are all islanders together. Our planet is itself but an island in the universe.

the diversity of species—a phrase which cannot fairly express the inestimable value for the future of the germ plasm of wild plants.

This is where the World Conservation Strategy comes in, and why it is so important. The world needs to keep as many of its plant resources as possible if we are to feed everyone now and in the future, and also have something left over in support of a little human dignity. All the world over, but especially in the islands, the best agricultural land needs to be carefully managed to give the highest yields. The marginal ag-

PART III

Plants For People

WITH HALF THE PEOPLE OF THE WORLD
either hungry or malnourished, economic botany
and human survival become ever more closely
linked. Even firewood to cook a simple meal is
prohibitively expensive in several countries and
shortages are likely to be far greater by the turn
of the century.

These seven chapters present a specialized
overview of economic botany throughout the
world, but mostly in the developing nations.
Here, to ensure survival, members of traditional
societies have learned how to cope with want,
and how to make the most of the simple foods
and fibers which grow near at hand. Furthermore, many of the most interesting plants are
found growing wild in remote surroundings, and
can be of importance not only to local people
but to all humanity.

The first chapter introduces areas of the
world that contain some of the most genetically
important wild species, the relatives of our main
food crops. Varieties that have arisen in cultivation, which botanists call cultivars, need period-
ical refreshment through crossbreeding with
their wild relatives. If we lose the wild fore-
bears, we will sooner or later lose the cultivars
to disease and to the erosion of productive vigor.
Then, in a picture essay, the reader can pause to
sample produce at local markets. This chapter
underscores an overwhelming dependence upon
plant sources of oils and fats, carbohydrates,
and protein by people in the developing nations,
and they are the world's majority.

Cellulose, the skeletal substance of most
plants, serves us all in many ways and its applications across the ages are the subject of an-

other chapter. An essay probes the shortage of firewood used by so many millions of people for cooking and heating.

Evolutionary development has transformed most green organisms into living chemical factories—sources of products that range from perfumes to poisons. These and other plant substances can alter the workings of the human body. The same organic chemicals that can heal, can also be employed in non-medical ways to gain power over others—through narcotic dependency, for instance. New facts about this, and another kind of "living death" are exam-

While the bounty of field and flower appear as if a gift of nature, human intervention through selective breeding is of great importance. The many qualities we desire in plants are found in wild species, but careful genetic manipulation enhances the value that nature imparts. Pictured below is a chromolithograph from nineteenth century America. Opposite; the Great Plains in harvest.

I am the Bread of Life

144

ined, Haiti's zombies, for instance. Toxins from narcotic vines and puffer fish are ritually administered to vicious or unreliable individuals, frequently ones who are held in contempt by the community. If not killed outright, the victims may be permanently impaired of mind, fit only for dog-work in the fields.

Two concluding chapters deal with healing herbs; seen first in light of their use in westernized medicine. Then the reader accompanies a mother and her sick child during their visit to the doctor—an herbalist of great repute in the East African country of Kenya. In times of ill-ness, most of the people in the world turn to such healers. The World Health Organization of the United Nations (WHO) has pledged its assistance to the development and improvement of traditional medicine, including midwifery. A notable example of this approach is the health delivery system of The People's Republic of China. It incorporates elements of both ancient and modern medical disciplines.

Moveable Feast

RICE ORIGINATED EITHER IN INDIA OR nearby areas of Asia. This ancient staple is shown growing on terraces in India, opposite, with stalks of American "Indian corn" standing in the foreground. Though *Zea mays* had its origins in the highlands of Mexico and Guatemala, it has spread around the world. Rice, too, is grown in both hemispheres. These are examples of "moveable feasts." Even the hot chili peppers so typical of the cuisine of India and Sri Lanka, and now grown widely in Asia, were introduced from Mexico and other American lands.

All our domesticated crops had their beginnings in plants taken from the wild. The evolution of the plant in its original habitat gave it the chemical or structural traits which we have exploited. It may come as a surprise to learn that nearly all of our major food crops were domesticated by Neolithic people from five thousand to ten thousand years ago. Barley and wheat were apparently first cultivated on the slopes of the Zagros mountains and their associated hills and valleys in what is now Iran and Iraq. There is archeological evidence that foxtail millets were amongst crops domesticated in Mexico's Tehuacán Valley seven thousand years ago. The Chinese have cultivated rice for five thousand years, with soybeans introduced a mere three millennia ago during the Chou dynasty.

In the first half of this century, the Soviet plant geneticist N.I. Vavilov drew attention to the large numbers of closely related plant species found in particular geographical locations. He suggested that these centers of genetic diversity might also be centers of evolutionary origin and dispersal. Although this view may be an oversimplification, it does provide a useful guide for economic botanists who seek new sources of genetic material for breeding.

Vavilov centers for wild types of beans are found in Mexico and Guatemala, for example, and corn also originated in the same area. The original peanuts, lima beans, tomatoes, peppers, cocoa, and potatoes came from South America, where many wild species still occur. Most wild brassicas—the cabbage group—appear in the Mediterranean Basin or adjacent Asia Minor, while Ethiopia possesses genetic reserves for sorghum and coffee.

Southwest and central Asia hold Vavilov centers for wheat, grapes, pears, beans, and barley. Buckwheat, millet, cinnamon, tea, and

Above; maize (Zea) is interspersed with the traditional rice crop on terraced plots in Southeast Asia. Opposite; the banana (Musa sapientum), a giant herb native to the area from India to New Guinea, is now grown in all tropical regions. This 1705 drawing from Suriname shows red bracts below the fruit unfolding to expose the male flowers.

P. Sluyter Sculp

Bushman woman exposes a marama bean root. Opposite; gourds hold edible marrow, disk-shaped bean root at upper right, and ostrich-shell canteens.

eties can have tragic consequences—witness the Irish Potato Famine. There was also an American corn blight of the 1970s, in which hybrids with one cytoplasmic factor died while others, even in the next field over, thrived with their slightly different genetic makeup. Today we understand that many domestic varieties must be periodically bred with wild plant stock to gain fresh resistance to disease. This is important, as chemical treatment is expensive and sometimes ineffective: and these agents are either unavailable or too expensive for many of the world's farmers. Even for agribusiness, genetic protection against disease makes good economic sense.

Wild species may have other attributes which some people find desirable. In fact, many people are willing to brave brambles and insect bites to gain the intense flavor of wild blackberries, raspberries, strawberries, and other choice morsels available only from the wild. The renowned Périgord truffle (*Tuber melanosporum*), perhaps the most desirable fungus, grows wild in woodlands of France and Italy and is the basis for regional industry.

Before Neolithic times everybody had to forage for food and familiarity with wild plants may have enabled our distant ancestors to develop the first primitive agriculture—where no one single variety dominated. In the Balkans, for instance, several kinds of wheat were in use by different groups of people at different times. Also, domestications were apparently made independently in different geographic localities. Scientists seeking new crop species fully realize the value of learning from people who depend on wild or locally cultivated plants. Botanical folklore can lead to scientific discoveries in both nutritional and medical areas.

This has occurred, for example, in the Great Indian Desert where local people eat a surprisingly wide range of wild plants, especially in times of famine. Several kinds of seeds supplement flour for bread. The women add fruits of a wild relative of squash to the pot containing domestic pearl millet. People here in Rajasthan State consume the nutritious fruits of wild and cultivated watermelons which grow spontaneously in the rainy season, and also sweet-

others came from China. Malaysia and Indonesia have coconut, clove, black pepper, nutmeg, bananas, and sugar cane. The Indian region has concentrations of wild orange, lemon, and tangerine, sugar cane and numerous wild species of rice, mango, and many legumes.

Northeastern India, the old Assam, is today subject to rapid development in the form of hydroelectric projects, new industries, growth of towns, and the extension of communications networks. All of these involve abrupt, large-scale, and irreversible changes to flora such as the wild mango, banana, citrus and other fruits whose races still grow there.

Why do we need to preserve at least some of these wild plant populations? The domesticated and selected varieties growing throughout the world are highly productive but often susceptible to disease. And in practical terms, excessive dependence on only one or a few domestic vari-

pulped fruits of various trees.

Hunters and foragers from antiquity, the Sen Bushmen of the northern Kalahari Desert survive by searching out more than one hundred wild species of plants and more than fifty kinds of animals. They realize that their reliance on such a wide variety of wild foodstuffs, scattered about a large area, can see them through years of drought. Wandering tribes of the Kalahari patiently search out wild relatives of the rattlesnake plant, or *Sansevieria*, a well known house plant. Several such fleshy plants draw moisture from the sands and accumulate precious water within their leathery leaves. They also collect wild melons, pods and other parts of several legumes, the bulbs of a plant in the lily family, and the dry fruits, seeds, and shells of the baobab. A euphorbia, a cattail, and the branched doum palm all add to the larder.

For these desert peoples life itself is one continuous hunt for these plants upon which their life depends. Some of their finds may one day profit all mankind. For instance, their marama bean (*Tylosema esculentum*) has more oil than our soybean and more protein than peanuts.

Governments and international agencies have in recent years ventured deeper and deeper into the wild to find new and underexploited sources of protein, oil, and starch. They are also discovering crops suitable for marginal lands and spent fields. Cultivation of some new and underexploited plants may conserve and even improve the soil. Because of some remarkable breeding successes, more and more scientists are focusing on wild or partly domesticated plants with desirable traits as the genetic resources for the future. Accounts of a few of the significant ones appear in the next section.

From Field to Table

IN TIMES OF ABUNDANCE, THE local markets of the developing world can be wonderful to see. They provide a rich variety of fruits and vegetables, grains, the various leaf plants, spices, and beverages. Many places, however, do not enjoy abundance. Most rural people eat grains and root crops, with relatively little meat or fish. While animal proteins are more nutritionally balanced than plant proteins, meat is more often than not a luxury, particularly in the tropics.

The "Green Revolution" of the 1960s and 70s emphasized new strains of rice, wheat, and other grains, with the exception of soybeans (a legume), for intensified cultivation. Legumes, however, have more protein than the cereal

Opposite, a young West African carries plantains, large bananas that are eaten cooked. Various kinds of bananas are cultivated throughout the tropics. Malaysia holds the largest number of wild species.

grains though they are seldom grown on as large a scale.

With larger acreage devoted to legumes, the incidence of such protein-deficiency diseases as Kwashiorkor can be reduced. Today, only a few of the eighteen thousand species are in common use: peanuts (groundnuts), soybeans, peas, lentils, pigeon peas, chick peas, mung beans, kidney beans, cowpeas, alfalfa, clovers, and vetches. Three main parts of the world consume beans to a great extent: eastern Asia (soy-

Basic Grains

Six of the world's major cereal crops—maize, rye, oats, wheat, rice, and barley—are represented here. From top to bottom, the spoons contain cornmeal (*Zea mays*), cracked rye (*Secale cereale*), oat groats (*Avena*), hominy grits (*Zea mays*), cracked wheat (*Triticum*), brown rice (*Oryza*), and barley (*Hordeum*). Only one of these grains, maize, originated in the New World.

Most ancient races of maize (called "corn" or "Indian corn" in North America) are found in Mexico, Peru, and Colombia. Cornmeal is used in tortillas, and hominy grits (spoon 4) are hulled and dried maize kernels prepared for boiling. *Zea* is now widely cultivated in all the Americas, southern Europe, West Africa, northern India, and parts of China.

Rye is usually grown for bread, and is most hardy in cooler north temperate areas, though its origin is southeastern Europe and Asia Minor. The Soviet Union is a major source and rye fields are also found in Europe, North America, and the Near East.

Groats are hulled, usually crushed, oats. The oat originated in the Near East, and was first grown in southern Europe as a pasture crop, later as a grain. Oats are mainly grown in Europe, North America, and the Soviet

Union, but a diversity of ecotypes makes oat growing possible from Alaska to South America.

Wheat is grown the world over except in the moist tropics. It was cultivated by the ancient Greeks, Persians, Romans, and Egyptians, and today the largest wheat-growing areas are in the Soviet Union,

Italy, France, the United States, Canada, Australia, and Argentina. It is also common in northern China, India, and parts of Africa. Wheat originated in the Near East and Asia Minor.

The unpolished grains of brown rice retain the embryo (germ) and the outer bran layer. Rice originated somewhere in Southeast Asia, a region devoted to its cultivation. It is also grown in Italy, the Iberian Peninsula, Brazil, West Africa, and the United States (California, Arkansas, Louisiana, and Texas).

Barley grains are used largely for soup, baby food, and in malting and brewing. The Vavilov center (center of species diversity) for barley is Ethiopia, and the center of origin may be in the Near East, with a secondary center in China. It is grown largely in Europe, the Soviet Union, and China.

This mural in a vegetable market in the People's Republic of China depicts fields where produce is cultivated. Special agricultural techniques permit refinement of such traditional dishes as Peking—or Beijing—duck.

beans), Latin America (common bean, *Phaseolus vulgaris*), and the Indian subcontinent (lentils, *Lens*), pigeon peas (*Cajanus*), chick peas (*Cicer*), and other pulses.

In addition to legumes and grains found in world markets, other botanical families which provide well-known food plants

are as follows:

Solanaceae (the family also of petunia, tobacco, belladonna, and mandrake)—potato, tomato, green peppers, chili peppers, eggplant (aubergine);

Rutaceae (the citrus fruit family)—orange, lemon, lime, grapefruit, tangerine, and "ugli" fruit (a

Beijing duck, commercially prepared, begins when the fowl is force fed special mash which builds up fatty tissues imparting fine flavor to the meat and crispness to the skin when cooked. As they roast in a special oven, the ducks are permeated by the smoke of wood from the Zizyphus jujuba tree, prized and jealously guarded by Chinese chefs.

cross between grapefruit and tangerine);

Cruciferae (the mustard family)—cabbage, Brussels sprouts, mustards, horseradish, radish, and turnip);

Umbelliferae (the carrot family)—carrot, celery, parsley, dill, and caraway seeds;

Rosaceae (the rose family)—apple, pear, strawberry, peach, plum, cherry, apricot.

To help solve the problems of protein deficiencies, it would be *more energy-efficient* to increase the supplies of vegetable proteins (such as legumes) than to increase the supply of animal protein, since the farm animal (e.g. cow, swine) is now the protein "middle man" between humans and the plant food to be consumed for protein. In effect, in many affluent countries great amounts of grain and leguminous fodder are required to produce meat for the table. Rural Africans often hunt and kill "bushmeat," the wild animals. It has been found that the wild species effectively convert plants into protein. Farming of wild species may be one way of assuring better nutrition for people of the tropics. In the jungle depths, river fish may be the most reliable source of high quality protein to supplement that from vegetable sources.

Versatile Cellulose

THE STRINGS THAT INQUISITIVE CHILDREN often learn to loosen and pull from stalks of grass are cellulose, the long-stranded molecules that make up the cell walls of most plants. It is often associated with various other chemical substances which impart a range of different properties.

Cellulose forms the silk of milkweed pods,

Photomicrograph of a tropical yam, double-exposure above, illuminates the configuration of its cellulose cell walls. The strength and flexibility of cellulose—the primary structural material of plants—are manifested in a Burmese shopkeeper's display, at right. He sells traditional crafts of a Shen tribe, whose people create an enormous variety of domestic containers and implements from bamboo, grasses, reed, cane, and other natural materials.

the parachute attached to dandelion seeds, and most of the structural tissue of herbs and woody plants. Its giant molecules are built from chains of smaller molecules of glucose, a sugar produced by photosynthesis. Each fiber in a boll of cotton, for instance, is almost pure cellulose, a single molecule perhaps ten thousand glucose units long. These filaments grow separated from each other, unlike the stem fibers of flax which grow together and must be teased apart.

Flax and reeds contain larger and tougher bundles of fibers than the small grasses that children tease into strands. Trees grow massive columns of bundled fibers. The layered structure of wood—the fibers impregnated with lignin, the tree's hardening substance—imparts great

strength and versatility. The list of uses for plant fibers is ancient and endless. New uses emerge all the time. For instance, regulation baseball bats are now made from the plentiful bamboo found in the Philippines. The maker glues numerous slats of planed bamboo into a solid piece roughly the size of the bat, and turns this blank to the desired shape on a lathe.

Like bamboo, rattan palms—woody vines—have played a major role in the life of the people of Southeast Asia, yielding the raw material for the rope used in bridges, houses, and other domestic and farm construction. Women weave the cane into baskets, mats, and furniture. World-wide trade in raw rattan amounts to $50 million annually. Indonesians cultivate several species, and the people of Taiwan seek substitutes for types made scarce in the wild by excessive harvest. The native yellow rattan, once plentiful

from the coast to the hilly interior of the island, is rarely encountered today. Perhaps the introduction of exotic thick-caned rattans from the South Pacific islands will ease Taiwan's short supply.

Kenaf is the name usually applied in the trade to the fiber obtained from the stems of two species of *Hibiscus*, *H. cannabinus* and *H. sabdariffa*, the latter also known as roselle. Thai jute and Deccan hemp are other commercial names. Eight-to-twelve foot herbs from tropical Africa, and naturalized in Asia, they provide an important substitute for jute with which their fibers are often mixed. It belongs to the mallow family, along with cotton and okra. Due to rapid growth, plants can be harvested after only four months. Some newspaper companies already include kenaf in their newsprint, and nearly equal amounts of kenaf and wood pulp may go into

According to the Old Testament story, a basket made of bullrushes like those being gathered near Cairo, at left, floated the infant Moses into the arms of Pharaoh's childless daughter. Processed another way, this giant sedge was the source of mankind's first paper, taking its name from the Greek word for the papyrus plant. Paper manufacture became a state monopoly. The city of Byblos, now in Lebanon, traded in paper with the Egyptians and provided the root of the English word Bible and other book oriented terms. To make paper, harvested stalks are split, treated in a vat, and pressed into sheets. Today this specialty craft provides material for artists who copy panels from tombs and elsewhere. Another fibrous plant, flax, supplied the miles of linen used to wrap noble mummies. Although cotton is virtually synonymous with Egypt today, it was actually not introduced until about 500 B.C. It too became a state monopoly, but arrived too late to contribute to Egypt's days of greatest glory. India was probably first to cultivate cotton as long as five thousand years ago and its growth in the New World is also of great antiquity—remains found in Mexico date from 3500 B.C.

future formulations. It is widely cultivated, the main producers being Thailand, India, and Pakistan. Thailand is virtually the only significant exporter and India is the single largest buyer followed by Japan, the United States, and the EEC. Overall, kenaf can help extend the world's supply of pulp wood. In the United States, where 97 percent of pulp comes from wood, the price escalated three hundred percent in the 1970s alone. Test plantings of kenaf have been made in Florida and Arizona, with an eye to supplying pulp to the large paper industry of the south.

Several species of *Agave* are cultivated for their leaf fibers, which provide 90 percent of the hard fibers of commerce. The most important is sisal (*Agave sisalana*), whose rosettes of tough, thick, spine-edged leaves contain cellulose fiber suitable for paper and cardboard. Mexico's state of Yucatán is a site of the cordage and rope industry for which sisal has traditionally been grown. Today the largest producer is Brazil, and extensive areas are under cultivation in East Africa. The prospect of diversification has spurred research on more efficient production methods and better land use in Mexico, where the government works closely with farmers.

Another fibrous agave plant of Mexico's Yucatán is henequen (*Agave fourcroydes*), the area's main export crop—one for which demand has dwindled. Stiff competition from Cuban, Asian, and African producers started the decline which was followed during World War II by a surge of development in synthetic fibers which took business away from Mexico. Although dozens of square miles are planted in henequen today, many fields lie fallow. Unless the market for henequen improves, the region will suffer a

A haystack is transported by ship on the Nile, below. The straw comes from grain fields, as it did in the days of Joseph and Moses, and it is still mixed with river mud for sun-baked bricks, at right, Egypt's traditional construction material. Although the bricks are water-soluable, rain is so rare in Egypt that buildings made of them may stand for decades.

Nepalese thatchers race fire in the Royal Shuklaphanta Wildlife Reserve, where for about a month each year they are allowed to gather grasses useful to their occupation as roof builders, opposite. Once the grasses have been harvested, fires are set to clean the remaining material, heavy stalks. These are woven into wall reinforcement for traditional daub-and-wattle construction. Since the reserve itself is surrounded by agricultural land, it is a virtual island of wild growth and the only local source of traditional construction materials. People drive their bullock carts as far as 35 miles to take part in the limited harvest of the reserve permitted by the government of Nepal.

great hardship, with the inevitable breakdown of traditional culture as people leave the land to seek work in the cities.

Examination of the henequen plant has laid the ground work for broadening its market. Suspecting that wealth might lie where the weight is, scientists first noted that a hundred pounds of leaves yielded less than five pounds of the long fibers traditionally considered to be of commercial value. The rest of the plant—including two pounds of waxy cuticle, nearly five pounds of short fibers, sixteen pounds of pulp, and seventy-three pounds of liquids—was discarded.

Hope for economic development may lie in the same chemistry that has created the prime fibers. For instance, a high-grade pulp for paper and cardboard is now made from henequen cellulose. Short fibers go into acoustic tiles and laminated roofing sheets. Given its substantial insulating qualities, this material may ultimately help to fill the economic niche formerly held by asbestos—a mineral fiber now known to cause lung cancer. Dry henequen pulp may find use as a filler in cattle feed. Also, leaf cellulose may be convertible into viscose rayon and other acetate products. Finally, synthetic steroids have been manufactured from the more than 40 tons of juices formerly discarded each year.

In spite of all this creative effort, the long

Below; refugees in a camp near Thailand's northern border learn to fashion large storage baskets out of bamboo strips.

164

As here in Africa, basket shapes for thatch roofs are common in the developing world. Machetes cut and shape the poles which are bound with vines and cord.

165

fibers of the henequen remain the principal moneymaker. Through selective breeding with the four or five other agaves native to Yucatán, geneticists hope to increase the plant's yield of commercial fiber. Similar work is also under way with sisal. Improvement of plant stock, often by growing entire plants from a few cells of a genetically superior individual, is of potential value for many lands. Such cloning can increase harvests without requiring additional land or placing heavier demands on soil already under cultivation.

Like sisal and henequen, the genus *Yucca* is in Agavaceae and grows in semi-arid regions of Mexico, the southwestern United States, and Latin America. Leaves and trunks of the woody species are used locally for building materials and fibers. High quality strands of *Yucca filifera* and *Y. carnerosana* go into belts, twine, brooms,

brushes, matting, and bags. In addition, scientists are now seeking industrial oils from the seeds, anticancer agents from the flowers, arthritis remedies from leaves, and paper from the trunks of certain species.

While herbaceous plants primarily supply the fibrous material for textiles, the woody ones are often used for shelter, furniture, and a myriad of other direct applications. Beyond this, in places like Sweden and Finland, trees are transformed into plastics, textiles, and other domestic

In some ways stronger than steel, flexible bamboo scaffolding covers a temple beside Burma's famed Shwe Dagon pagoda at Rangoon. Right; in Calcutta, tied bamboo scaffolding undulates up the side of new construction.

166

materials. These countries also produce great amounts of paper and plywood. Furthermore, their forest management is among the best. The wise course might be to commission northern lands to provide more pulp for paper, and for some other uses. The boreal forests of Canada, Scandinavia, and the Soviet Union can regenerate themselves easily. It happens because northern softwood forests are nature's monocultures. Nations of the moist tropics might specialize in other fields, augmenting their cash income through managed harvest of their prized hardwoods—an approach recommended in the chapter on Tropical Rain Forests.

The increasing global demand for lumber and paper products is posing an ecological challenge everywhere, but primarily in the tropics. The rain forests of eastern Australia, along the coast of Queensland and New South Wales, are threatened by economic exploitation. In Chile and Argentina the monkey puzzle tree is being harvested for plywood. Stands of the enormous Chilean larch are dwindling and the remaining forests of southern Chile have become potential wood pulp for foreign—generally Japanese—paper companies.

The news is spreading throughout the less developed world that tree farms are profitable, especially with the faster growing types. For the many nations unable to afford the petroleum that they want, self-sufficiency in tree production can also mean self-sufficiency in fuel. Planning wisely for the disposition of precious forests and other lands capable of producing marketable cellulose can bring nearer the day of sustainable prosperity.

Woodsman at India's southern city of Trivandrum cuts a bamboo thicket's top shoots. He has ascended on small branches that sprout from the joints, or nodes, of the plant's hollow segments. Below; trimmed bamboo poles are carted to market by a team of zebu cattle, prized for their horns fused at the tip.

Firewood

INDIA HAS ENOUGH FOOD BUT NOT ENOUGH fuel to cook it, and in many lands the firewood needed to cook a meal is more expensive than the food itself. As the distinguished botanist T.N. Khoshoo, now Secretary for the Environment in the Indian government, has written:

> . . . apart from other basic needs like food, shelter, clothing, and medication, energy for cooking food and heating is also a fundamental and basic human need which has to be guaranteed by the state.

According to the Food and Agriculture Organization of the United Nations (FAO), three people out of every four in the developing countries use traditional fuels for their daily domestic energy. For more than a quarter of the human race, even minimum needs for cooking and heating cannot be met; one hundred million of these people live in areas where woodlands have been depleted. With fuel unavailable from the usual sources, people in parts of India, the Middle East, and Africa cut down trees in parks, gardens, and wildlife reserves. They also burn the animal dung that is sorely needed to fertilize crops. Another billion people live off the countryside where wild plants are being rapidly depleted.

People living in the dry zones of Asia and Africa experience grave difficulty in obtaining the life-supporting product daily. Because plants grow slowly in arid regions and produce only limited amounts of greenery, the ecological value of each plant is particularly high. Trees, bushes, and other plants slow the movement of sand dunes, guard soil from erosion, conserve moisture, and also provide people with food, clothing, shelter, and fodder for livestock. As in any setting, though, shrubs, trees and grasses are also in demand for firewood.

For domestic use, most wood is consumed in low-efficiency traditional cooking stoves and open pits; and often wastefully burned while still moist. Thus the time and effort needed to obtain the wood go up in smoke with the squandered fuel. In the mountains of Nepal, for example, a person may now spend an entire day gathering the same modest supply of firewood previously collected in an hour. Some West Africans must walk 15 miles to locate sticks to burn. In Niamey, Niger, manual laborers spend a quarter of their income on wood. Similar conditions prevail across the Sahel and in the Andes.

Her fuel shop piled to the rafters with branches and split logs, a proprietress of Canton, China, examines her accounts, opposite. Some wood comes from city forests, tree-lined avenues which are carefully tended, the prunings conserved for firewood. Below; an Ethiopian woman prepares her cookfire, a daily ritual in the many lands where wood fires provide heat for cooking.

In most countries fuel is obtained not from forests, but from drier woodlands such as Africa's savannas, from brushwood growth, and from local woodlots. With increased population pressure in the tropical forests, virgin forests are being cut into more deeply, especially as other sources of supply become exhausted.

In spite of rapid deforestation in the humid tropics, the overall shortage of firewood is not yet critical. Tropical conditions produce high rates of photosynthesis and high yields of living matter. Thus, the two hundred million people living in or near tropical forests (as opposed to the one billion people living in farm villages and cities in humid tropical zones) are sometimes in a position to sell wood and charcoal to faraway city markets.

Firewood, especially in the form of charcoal, is increasingly an article of international trade. Kenya exports charcoal to the Arabian Gulf region and Suriname sends it to northern Europe, to name only a few markets. Although this traffic produces needed revenue for the seller, it can put undue pressure on forests and other areas which are being fully exploited for local use. There are numerous reasons for charcoal's popularity: it is easier and cheaper to transport than wood; its heat is steady and concentrated; it is relatively smokeless; and it can be extinguished and relighted. On the other hand, the burning of charcoal produces only half the energy originally in the wood, the rest lost as the wood was smoldered into pure carbon inside the charcoal kiln.

Even greater quantities of wood are wasted in the technique of shifting agriculture practiced by forest farmers. Trees are cut and burned on the spot to clear the land. The ashes contain soluble potash which is dissolved by the first rain. Most of the essential plant food washes away, but the soil does soak up a little. To obtain this poor-man's fertilizer, forest farmers burn as many as six hundred and fifty million cubic yards of timber each year.

A recent FAO assessment outlines the particular areas involved in the firewood crisis, a global dilemma made up of numerous regional, national, and local problems. First are the driest

regions, those having very few resources and low populations: the Sahara desert, Somalia, South West Africa (Namibia), the entire Arabian Peninsula, Iran, and the Chilean Atacama. Second are areas of scarcity and higher population density: El Salvador, Jamaica, Haiti, northeastern Brazil, central Ethiopia, central Turkey, and Nepal. The third category, those countries with a current deficit, includes central Mexico, Cuba, southern Guatemala, the Dominican Republic, Trinidad, coastal Peru, Niger, central and western Madagascar, parts of Angola, the middle Philippine Islands, Java, and almost all of India.

To the three categories above, we add a fourth—areas for which a firewood deficit is imminent. Uruguay, coastal North Africa, Mali, central Sudan, southern Chad, Zimbabwe, Tanzania, Viet Nam, and northern Sumatra are all likely to gain firsthand experience of the firewood crisis before the turn of the century.

Additional wood could be made available for cooking if such materials as bamboo, straw, and the stalks of sorghum and millet were em-

Charcoal's life cycle: a kiln in the Mexican jungle spews steam and chemicals driven from wood by heat. Beehive kiln, top, in Pakistan has been dismantled to reach the charcoal. Right; Middle Eastern blacksmiths work metal in a bellows forge.

ployed by people in the arid zones for construction and fencing. Rather than allowing livestock to graze on any woody plant in sight, cattle can be confined and fed such plants as the spineless cactus and atriplex, reducing the pressure of grazing on fuel plants and permitting them to gain strength for reproduction.

Plantings of every size from woodlot to forest should be established to serve farmers and townspeople. Protected tree farms of joint ownership, woodlots, create an economic resource while reducing stress on the habitat. They also help to lower the cost of wood and the necessity of packing it in from distant sources. This approach embodies the ideal of *conservation for production* which should be the basis of economic planning for the developing world.

Turning ideals into achievement, however, often requires a great expenditure of time, effort, and money. In the view of the World Bank, up to ninety-six thousand square miles of trees must be planted by the year 2000 just to match demand. This estimate optimistically assumes the use of efficient wood stoves, production of methane fuel from the fermentation of organic matter, and the widespread adoption of solar cookers. Even so, if current rates of afforestation around the world are not exceeded, only a tenth of the required new plantings will exist by the end of the century. Tree planting in Africa's Sahel may need to increase fifty-fold if projected demand for the year 2000 is to be met.

In many locales, the most promising approach to afforestation is to identify appropriate local trees and bushes and to speed their domestication. Fortunately, many underexploited spe-

cies are being discovered in the wild. Some of these grow so fast, in fact, that they can become weeds if not controlled. Proper controls should be established before their introduction.

It is also particularly fortunate that a wide spectrum of woody plants can be grown in arid zones, those regions of greatest need, and that many of these species—various kinds of acacia, eucalyptus, and the neem (*Azadirachta indica*)—are suitable for plantations. The pigeon pea (*Cajanus cajan*) is a firewood no forester would recognize, for it is a food crop with a woody stem. With this relative of the black-eyed pea, people can feed themselves and cook their food from

the same plant.

Coordinated programs can facilitate the distribution of seeds and seedlings to those who will tend the plantings. Young trees are already provided free or at low cost by government agencies of several countries. Local woodlots and plantations are part of an integrated program. This global operation's first priority, however, involves the rescue of promising wild firewood species now threatened by intensive cutting, often in the areas where fuel shortages are greatest. These are the breeding stocks, and the areas where they occur deserve protection as international biological treasures.

Some prime genetic materials exist in restricted habitats, fragile life zones at the edge of extinction. Cooperating countries and international organizations can work together to conserve these Vavilov centers—pockets of species diversity and genetic vigor uniquely valuable to breeders. The accelerated planting of firewood plantations for people who live near these priceless genetic repositories is a first step toward prevention of loss. And as elsewhere, these plantings will also create employment for producers and distributors of the wood, and provide shade, shelter for people and wildlife, as well as beauty, food, soil stability, and watershed amenities for the inhabitants.

Reforestation should be a primary conservation goal for this century and beyond. Much depends on success in easing the global crisis. Proper nurture and use of woody species and their habitats lies at the heart of many, if not most, of the efforts designed to safeguard the future of our green world.

With laden camel, a young Afghan searches for bits of kindling and firewood in a desert area of Afghanistan. Below; an Indian merchant has neatly stacked dried animal dung (better used to fertilize fields) to sell for fuel.

Chemical Cornucopia

The milky latex from the succulent stem of the Euphorbia *below contains a toxic resin which may cause paralysis. Nonetheless, with proper attention to dosage, it has been used effectively in remedies for cancers among people of southern and eastern Africa. Like the sap of another euphorb, the castor oil plant, it is also often used as a purgative. Here, the plant's vascular pressure creates a bubble from the latex.*

FOR THE MOST PART, ANIMALS PROTECT AND sustain themselves through locomotion, while plants build up a variety of chemical substances for their stationary defense. Physiology has much to do with how plants and animals obtain food and the amounts of energy needed to secure their nutrition. Plants are adapted to consume sunshine for their energy and animals eat either plants or animals that eat plants. Thus light is the basic energy source for all living things on earth, but only plants are able to transform it into discrete substances, chemicals of varying complexity. Their dependence on available sunshine and water generally forces plants to live within a stricter energy budget than animals. To make the most of their method of survival, to use the energy available to them with the greatest efficiency, plants have evolved into chemical factories—sources of all sorts of raw and finished materials which people appropriate.

Previous chapters have mentioned cellulose, the basic organic chemical from plants, and lignin, the material that impregnates the fibers and hardens them in wood and some other plant tissues. Oils of many kinds are made by plants, from essential oils such as the attar of rose to the fatty oils such as cocoa butter. Sweet substances abound, from sugar cane and beet to palm jaggery and maple syrup. The dark and bitter seeds of the cacao pod are dispersed through a white pulp of intense sweetness, and some plants have chemicals which taste sweet but contain no sugar.

The bitter parts of plants often contain alkaloids, complex nitrogenous compounds. Many of these produce marked physiological responses in higher animals and are thus of interest to medical researchers. The value of alkaloids to the plants themselves is debatable, but such potent chemical substances may be part of plants' defense mechanisms—their means of discouraging browsing animals. Thick skin or bark, spines, and thorns appear to serve a similar function. Common alkaloids in daily use all over the world include caffeine in tea and coffee and chocolate's theobromine—chemically almost identical to caffeine. Opiates from poppies and other narcotic producers, quinine from the bark of cinchona trees, digitalis from foxglove, and numerous others alkaloids find medical use.

The white milk which drips from wounds on plants is called latex. *Hevea brasiliensis* and sapodilla (*Manilkara zapota*) plants yield a milk rich in chemicals called elastomers—chemicals which stretch. The former is known as rubber and the latter as chiclé—the substance from which chewing gum originated. The boom in rubber as a major plantation crop in the late nineteenth and early twentieth centuries was due to its use in bicycle and automobile tires. During World War II, after Pearl Harbor, supplies of Asian rubber were cut off. The search for alternate sources immediately gained high priority. According to Charles M. Wilson in his remarkable book from the war years, *Trees and Test Tubes, The Story of Rubber*, fuel rationing and the imposition of a strict 30 mile-per-hour speed limit in the United States were primarily to save rubber, not petroleum.

Of the hundreds of plants that produce rubber, only a dozen or so are of commercial importance. With the exception of the Para rubber tree (*Hevea brasiliensis*), the most promising natural source of rubber today is *Parthenium argentatum*, the desert plant commonly known as guayule. Yet neoprene and other synthetics so captured the public imagination after the war that this natural product was almost totally neglected. A member of the Compositae or daisy family, guayule grows wild in northern Mexico and the Big Bend area of southwestern Texas. Both hevea and guayule rubbers are identical in physical and chemical properties, the primary difference being that the latex hardens in the bark and wood of guayule stems. It was made into bouncy spheres by the ancient Aztec Indians for their ritual "basketball" games—with the latex extracted from the plants through a communal chewing of the stems.

In 1980, the United States imported a quarter of the world's rubber supply from hevea plantations, most located in Indonesia, Malaysia, Thailand, Liberia, and Sri Lanka. At the same time, Mexico imported nearly 90 percent of its natural rubber.

Research with superior breeding stock may help make guayule an economical crop. Mexico and the United States are working along these lines and pursuing improved techniques for the extraction of hardened latex. The federal governments of both countries, tire makers, and agricultural companies are seeking ways to commercialize guayule, a plant whose cultivation could be useful for arid countries. Further research and development is required, however, before guayule can enter world markets.

Natural rubber stands a good chance of holding its own in international trade, partly because latex is still essential for certain pneumatic tires. Those on airplanes are nearly pure natural rubber, as are many radials on large trucks since natural rubber can stand much higher temperatures than synthetics before disintegrating. In a form often called plantation crepe, Para rubber gives its famous bounce to fine casual and sports footware.

The economics of rubber seem to be changing since petroleum, the chemical feedstock of synthetics, has grown far more expensive. As researchers look for substitutes, economics is dictating what appears to be a serious movement back to natural products. In terms of fuel for transport, Rudolf Diesel wrote in 1911:

> *The diesel engine can be fed with vegetable oils and would help considerably in the development of agriculture of the countries which will use it. This may appear a futuristic dream but I can predict with great conviction that this use of the diesel engine may in the future be of great importance.*

Oils in latex from plants in the daisy, spurge, and bean families are being investigated for use as alcohol fuel substitutes. A winner of the Nobel Prize for his investigation of the chemistry of photosynthesis, Professor Melvin Calvin of the University of California at Berkeley, has pursued the fuel oil potential of *Euphorbia lathyris*, called gopher weed in the United States where it is fairly common. This relative of spurges and poinsettias came from Europe where it is known as myrtle spurge and mole plant. Professor Calvin has obtained a yield from crushed plants equivalent to several barrels of hydrocarbons an acre, which could perhaps be improved through genetic selection of plants of high yield. *E. lathyris* produces a blend of carbon compounds known as terpene trimers. These, in turn, produce some substances almost identical to those derived from naphtha which cost approximately $50 per barrel in 1983.

Even more spectacular is the copaiba of Amazonian Brazil, *Copaifera lansdorfii*, a leguminous tree whose trunk can be tapped to provide an oily sap used by local people as an anthelmintic, as well as an ointment for treating cuts, and a fuel for lamps. The trunk is drilled to tap the oil, and when the flow ceases the hole is closed with a wooden and clay plug. A diesel truck with no modification to the engine has run on the unprocessed sap.

Further along toward commercialization is oil from the jojoba bush of American deserts, limited today to the cosmetics market because of its high price. Jojoba was first mentioned in print in 1789—the year in which the French Revolution began—in a book published in Venice, Italy. In his *Storia Della California*, the Mexican historian Francisco J. Clavijero reported that the jojoba shrub was held in great esteem by the Indians of Baja California, who ate its fruit and used its oil to treat cancer and kidney disorders. A member of the boxwood family, Buxaceae, jojoba is known to science as *Simmondsia chinensis*, although it does not occur in China as suggested by the species name. Its home ground is the Sonora Desert of Mexico, Arizona, and southern California, where summer

temperatures can reach 115° F in the shade.

An evergreen shrub with leathery leaves, a deep root system, and a tolerance for saline and alkaline soils, jojoba can grow to a height of more than ten feet. Unique in the Plant Kingdom, its olive-sized seeds contain a liquid wax rather than the commonly found fat or oil. Chemically, it is identical to oil of the sperm whale. By hydrogenation, the catalytic process that creates margarine from vegetable oil, the jojoba product emerges as a hard white solid.

Until recently the whale's liquid wax has had no peer as a lubricant in automobile transmissions and other precision machinery subjected to high pressure and temperature. Jojoba seeds contain half their weight in a raw material which needs little refining to be used as a lubricant. In technical terms, its viscosity, flash point, and fire point resemble those of sperm oil. It is without odor and taste, and is not damaged when heated repeatedly to high temperatures. This means that jojoba is probably suitable for everything done by sperm oil and much more. This includes an amazing potential for the cosmetics industry as a component of hair oil, soap, shampoo, face creams, moisturizers, sunscreen compounds, and various conditioners.

Both cane and beets supply sugar, probably the most popular plant sweetener on earth. Bubbling sugar syrup from the vat stirred by an Indian workman, below, may be cast into hard sugar loaves. The Arabic word for sugar cane, gand, *is the source of the English word "candy."*

Saffron—The Golden Spice

Saffron, the costliest spice, has long been called edible gold. This is partly due to the hand labor required for its extraction, the spice itself being the dried red stigmas, or female organs of the *Crocus sativus*, and four thousand stigmas produce only one ounce of the spice. Because the purple flower is sterile all of the saffron crocuses form a clone. As a result of the division of the corm, propagation of the species is labor intensive.

Historically, the Levant was a center of saffron trade, in a time when the lovely golden yellow of the stigmas was employed as a dye for expensive cloth. Crusaders brought bulbs home with them from the Holy Lands. Today Spain and Turkey, where the flower is harvested before the stigma is removed, are the main producers.

Chemicals in the stigmas impart a delicate fragrance and an attractive tinge to rice dishes, including Spanish paella. They also find use in specialty liqueurs and tonics. For centuries saffron has been considered a panacea, a general stimulant, and even an aphrodisiac. Francis Bacon observed, "It maketh the English sprightly." Whatever its clinical attributes, saffron has long been the subject of adulteration. So prized was it in the Middle Ages, that people caught tampering with the pure product were often executed.

Warning:

Essential oils and other chemicals in spices and other plant products are often poisonous if consumed in large quantities. It is dangerous, possibly even fatal, to employ self-dosage or to experiment upon one's self with local plants. Many herb teas have not been scientifically investigated. Furthermore, some plants displayed during holidays, including poinsettia and mistletoe, are toxic. Laburnum pods and seeds are poison. Children should not put any parts of plants in their mouths or near their eyes. Most pediatricians have charts of toxic plants. Large cities often have poison control centers, where children can be taken in an emergency.

With all their oily, waxy, and aromatic juices, saps, and oozes, plants are natural sources of new insecticides and pesticides, additions to the traditional arsenal of tobacco derivatives, pyrethrum and rotenone plant products. Scientists of the University of California at Berkeley are investigating chemicals produced by an African bugleweed (*Ajuga remota*) of the mint family. Its chemical extracts interfere with molting of cotton pests such as the pink bollworm and the fall armyworm. The plant is used by African herbalists to combat high blood pressure and malaria, but the Berkeley scientists found that it will literally cause the caterpillars to starve inside their own repeatedly molted skins. Usually the old skins break open and are shed, but not after applications of *Ajuga*. Ways are now being investigated to obtain larger amounts of the bugleweed chemical, ajugarin.

The neem tree (*Azadirachta indica*), a member of the mahogany wood family, is also showing promise as a source of natural insecticide. Native to India and most common in the dry Deccan forest, it yields timber, gum, astringent bark, and various medicines. Indian farmers and merchants add neem to stored grain to prevent insect damage. Even when ground into flour or meal, and thus eaten with the grain, neem has not been found harmful to people. From its roots to its trunk, bark, leaves, flowers, fruit, and seeds, almost every part of the plant is put to use in the countries where it is grown.

Neem's natural insecticides, mainly azadirachtin and related compounds, are concentrated in the seed and released by grinding. Applied in suspension to a crop, the compounds penetrate each plant's vascular system and are thereby spread, repelling attacking insects. Fortunately, azadirachtin and other repellent compounds reach into newly developing leaves and branches.

Neem deters a truly amazing catalog of insect pests: desert locust, tobacco budworm, cotton stainer, and fall armyworm; beetle and moth pests of stored grain. Japanese beetles will not even touch a treated plant. There are many others including the lowly housefly. Trials are being carried out with neem constituents for pest control in the United States.

Tests are also underway to investigate neem's ability to become established in hot dry regions without native timber. Neem may also improve the soil. Already grown extensively as a shade and lumber tree in arid and semi-arid

The South African Aloe *at right is the source of active ingredients for tonics and cathartics. The sap of other species soothes burns.*

tropical areas of Africa and other regions of the Old World, it is being introduced, though quite slowly, into the Western Hemisphere—primarily in Puerto Rico, the Virgin Islands, and Haiti—where it has been a street tree since the 1960s. Its world cultivation is somewhat uneven; in 1981 there were approximately 25 million neem trees in India, and only two in the United States, both in Miami.

A chemical from endod (*Phytolacca dodecandra*), an African plant related to the American pokeweed, may help free the human race from one of its great afflictions. As many as two hundred million people harbor the tiny blood flukes which cause schistosomiasis or bilharzia. These parasites are passed to unaffected people who bathe in the same streams used by those with the disease. Water snails harbor the flatworms which gain entrance to the human body by burrowing into the skin. Distress, damage to internal organs, and death may occur. The ability of the endod plant to kill water snails was noted by scientists in Ethiopia. To wash their clothes, rural people go to nearby streams and use the endod, a pokeberry, as their detergent. Water downstream is free of snails. Local cultivation could supply the chemical to communities unable to buy commercial pesticides.

Throughout the world, in rural areas as well as in large cities, many potentially useful plant chemicals either poison or addict their users. Other substances, those derived from mushrooms, find ethnic or religious use, as with various hallucinogens employed for tribal rituals in the Americas and elsewhere. The coca plant (*Erythroxylum coca*) from Andean nations and the opium poppy (*Papaver somniferum*), have an illegal following in virtually every country on earth. Though the course of chemical addiction has been known for thousands of years, only since the 1970s have scientists begun to understand and to attempt detailed explanations of the various forms of chemical intoxication. The following pages reveal surprising facts about the non-medical uses of plant substances, often toxic, that have become part of the culture of two societies.

JOSEPHINE H. JOY.

Zombies

The meaning of the voodoo religious cult is as remote and elusive to the tourists who attend ceremonies staged at Port-au-Prince hotels as it must have been to the earliest white ethnographers. Yet, as recently as 1983, a remarkable breakthrough towards the scientific understanding of one of the secret voodoo phenomena, zombism, was achieved by Dr. Wade Davis, a Harvard University ethnobotanist.

A theme that has pervaded the lore and physical reality of voodoo for generations is that certain practitioners apparently possess magic to raise the dead. Indeed, the undead were occasionally seen.

They wandered about, or labored in the fields, cutting and weeding the sugar cane with slow and deliberate motions suggestive of being in a stupor or trance, as if the soul had been removed from the body: the living dead.

It is now recognized that the communities of voodoo priests (*houngans*) include some specialists in "black magic" called *bocors* who belong to the secret society of the *Zobops*, whose *loas* (spirits) can be invoked for harmful practices. A zobop is most often consulted, as noted by Dr. Davis, "when individual passions such as love, hate or jealousy express themselves." Alternatively, zombification may be dealt to a person whom a family member or the vil-

lage community at large wishes to punish.

The bocor is able to induce in the victim a deep coma. The body's vital signs are arrested to the point where the victim appears "dead." This is done by surreptitiously administering, to the skin of the intended victim, a powdered mix of toad and puffer fish poisons and plants with stinging hairs. After an "accidental" brushing of the poison, it is absorbed into the skin and the victim starts to feel the effects. The zombie poison contains two species of plants which are hallucinogenic and can cause amnesia: *Datura metel* and *Datura stramonium*, both of which are called "zombie cucumber." Another species used in the pow-

der is *Mucuna pruriens*, a plant with stinging hairs and one which "contains psychomimetic constituents and may have hallucinogenic activity."

The victim may recognize the early effects and repent in time to receive an antidote from the priest. In other cases, the poison continues to act until the victim is certified dead and buried in a coffin. Obviously, no autopsy or embalming is permitted by the person who arranges burial, often one very close to "the deceased" and a collaborator with the zobop.

Under cover of night and usually well within two days of burial, the person is disinterred and revived. In Hector Hyppolite's painting, *The Zombies*, is seen the cemetery with the black crosses and the death's head emblems of Baron Samedi. Antidote from the bottle in the bocor's hand has resurrected the two zombies, who are roped together for restraint during the violent spasms always exhibited by an awakening zombie. These drugs and their dosage have interesting potential for the fields of psychiatry and space medicine. The most mystifying aspect of zombification, from the medical point of view, is the great degree of expertise required to limit dosage of the poison, and to mix the correct amounts of plant and animal materials into it, so as to approach death, but to fall short—just this side of eternity.

Blue Dreams

The Correguaje Indians along the Caquetá and Orteguaza rivers in southern Colombia prepare a narcotic drink called *yage*, signifying "blue dream", from the bark of the *Banisteriopsis caapi* vine, a tropical liana. Hallucinations in vivid color are experienced, and at least 50 other tribes in northern South America utilize the "vision vine" for this purpose, often cultivating it in dooryards.

From the feathers inserted into the earlobes of this Correguaje are strung the winged fruits (left) and flower clusters (right) of the liana. Due to the high consumption, now approaching mass addiction, the *Banisteriopsis* (commemorating Reverend John Banister, a missionary) has become very rare in the region inhabited by this tribe. The psychoactive chemicals causing the *caapi* intoxication appear to be alkaloids.

Dilute tinctures of yage are prescribed by Colombian homeopathic doctors for a variety of conditions: memory loss, lunacy, delirium, sea-sickness, frontal headache, epilepsy, paralysis of tongue and side of body, cardiac stimulus, liver ailments, polyuria, and as an aphrodisiac. Such local use may indicate the presence of a variety of active chemicals in the plant which, if scientifically analyzed and investigated, might yield medicines of value.

Medicine's Hidden Herbs

Wood and bark of Quassia amara, *shown in fruit below, yield a bitter substance long used in tonics. Opposite, a turn-of-the-century advertisement promotes a "cure" for various respiratory disorders. Cherry extracts have long been valued for their soothing and bracing qualities and remain popular in cough syrups and candies.*

Norman R. Farnsworth, indicates that in 1980 a quarter of the prescriptions dispensed from community pharmacies in the United States contained active principles from the higher plants—these chemicals numbering about 100. The survey does not include medications used at hospitals, or prescriptions including the so-called wonder drugs. Penicillin, streptomycin, and the others usually are fungal extracts. Though part of the Plant World, fungi are not considered part of the Plant Kingdom proper—rather a separate domain (see page 36).

These 100 most important medicines from green plants are derived from only 41 species, with several of the alkaloid-producing plants contributing more than one compound. Opium poppies contain a number of useful alkaloids including morphine, codeine, and papaverine. The genus *Strychnos* produces the medically useful compounds strychnine and the curare of arrow-poison fame—without which modern surgery would not be possible. The heart and other organs are relaxed with curare, while life is sustained through the use of machines. From the mandrake (*Mandragora officinarum*) comes scopolamine, the "truth serum" of World War II espionage fame, now a standard pre-operative medication. The heart stimulant digitalis comes from a wild foxglove of Europe, and steroids from wild yams provide precursors, or raw materials, for hydrocortisone and birth control pills. One of the earliest plant drugs to be synthesized was aspirin, in 1899, in which a substance from willow bark provided a chemical blueprint for the synthesis.

Aspirin, a "wonder drug" in its own right, was followed during World War II by penicillin, first of the antibiotics, an extract from a common fungal mold. Since then, literally dozens of healing substances extracted from molds of the soil and from other sources have been cultured in vast quantities in special nutrient solutions. The knowledge of culture mediums and their industrial exploitation represents one of the great advances in medical and nutritional technology. Vitamin C, ascorbic acid, once came primarily from lemons, black currants, acerola, roses, and other green plants. Today, microscopic fungi grown in vats provide most of our ascorbic acid.

A recent wonder drug, Cyclosporin A, was first cultured from a soil sample gathered in Madison, Wisconsin, by an employee of a Swiss pharmaceutical company. The effect of the ex-

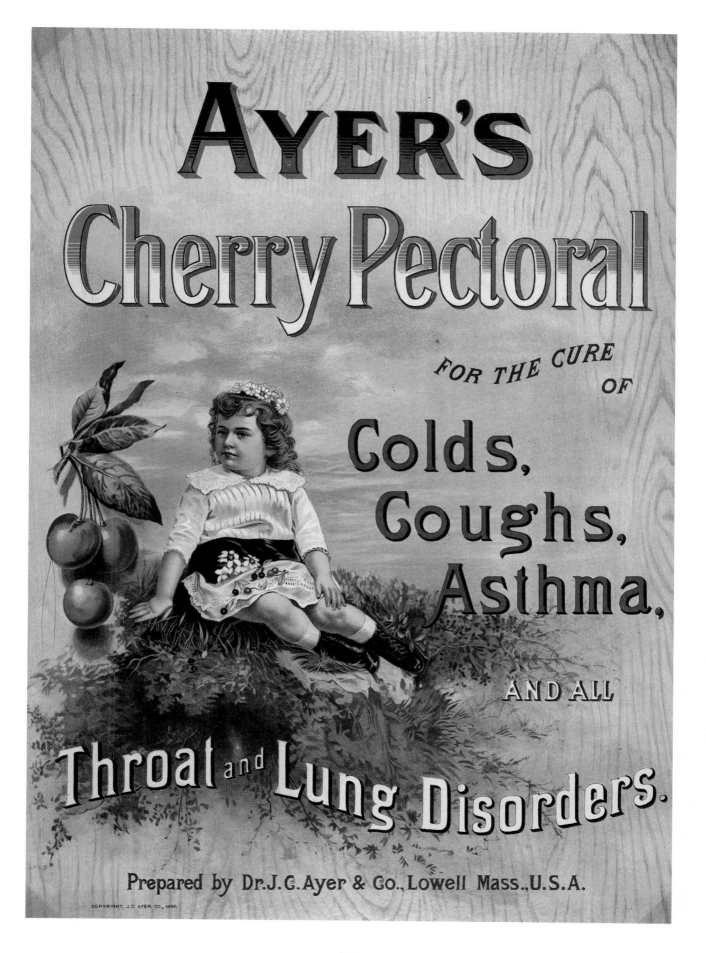

Flower That Cures Cancers

The patient was a 49-year-old machinist. For nine months he had suffered from generalized Hodgkin's disease . . .

When he came to Lilly's Clinical Research Laboratories, he carried a ten-inch tumor in his chest . . . on March 2, 1960, he received his first injection of Lilly's new anticancer product, Velban. Within four days his pain had disappeared. In a week he was walking. After 12 days his tumor began to shrink . . . in four (months), the tumor was gone.

These words come from a copy of a 1961 "Lilly Review," house newspaper for Eli Lilly and Company, and seem almost too miraculous to be true. But so was an announcement in 1983 that for the first time in history the cure rate for cancer had risen above fifty percent in the United States. The statistic is quite true, as is the above quote. Until recently, though, those who knew best about the effect of the chemicals vinblastine and vincristine were cancer patients who could be treated either at the doctor's office or in a hospital for a short period each month. At other times the person undergoing chemotherapy could resume a near-normal life as treatment progressed.

The chemicals behind the cures began as alkaloids manufactured naturally within a small flowering plant, *Catharanthus roseus*. Commonly known as the rosy periwinkle or Madagascar periwinkle, the flower has been widely introduced beyond its original island. Velban, the Lilly trade name for vinblastine, derives from an abundant compound within the plant's leaves. Another cancer drug from the same plant, Oncovin, begins as a very scarce alkaloid called vincristine. Fortunately, due to a process developed in Hungary, the more abundant vinblastine can be transformed in the laboratory to vincristine.

The discovery of both of these active substances, now produced commercially by Lilly, can be traced to a large-scale effort of the United States National Cancer Institute in the 1950s to screen plants for anti-cancer properties. Industry help was enlisted. Scientists at Eli Lilly were intrigued by folklore from Africa and elsewhere which associated the rosy periwinkle with treatment of diabetes: scientists separated more than eighty alkaloids from periwinkle leaves but found no medicine for the disease. Vinblastine, however, reduced the white cell content of the blood. Runaway production of white blood cells is one characteristic of leukemia.

Indeed, without vincristine therapy, there would be no cures of childhood leukemia. Also, in combination with other medicines, vinblastine therapy achieves a cure rate of 80 percent in cancer of the testicle, formerly fatal in most cases.

The periwinkle alkaloids apparently prevent cell division in animals and man. The wild growth of tumor cells and the unchecked division of white blood cells are reduced through chemotherapy. Fortunately, chemicals from the periwinkle work in concert with other agents, and even alone often have the power to initiate a remission. Then the other medicines can work more effectively to destroy the original cancerous growth and also stop the spread of cancer cells

from the active malignancy. Coupled with strict schedules of injections, called protocols, the various formulas and dosages are especially effective.

While cures for cancer are a blessing, further knowledge of plants may aid in prevention of the disease. Recent government announcements in the United States suggest that changes in diet and smoking and drinking habits can reduce chances for the onset of cancer. In years ahead we may well possess the ounce of prevention as well as the pound of cure. Cancer may one day become as rare as polio and smallpox.

Two alkaloids derived from Madagascar periwinkles—shown growing in Texas, opposite—play a major role in cancer chemotherapy. They contribute to the high rates of cure for previously fatal childhood leukemia, and for some other malignancies.

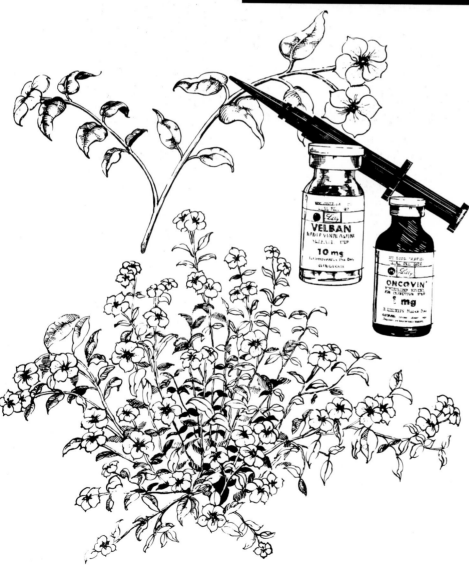

This well-stocked modern herbal pharmacy in India reflects the important role of traditional medicine in India's health care delivery system. Colchicine, an alkaloid effective in the treatment of gout, comes from the seeds of the attractive glory lily (Gloriosa superba) at right. The flower's popularity for bouquets has produced a shortage of the seeds, and cultivation of the lily on "pharmaceutical farms" has been recommended.

perimental extract was strong, but this particular organism could not be cultured in bulk. *Colypocladium inflatum*, a more adaptable species, could. It turned up in a Norwegian fjord. The drug suppresses rejection of newly implanted organs. Liver transplants have become far safer due to the drug, which may also help fight malaria.

The well is indeed not dry. Norman Farnsworth notes that of five thousand species of flowering plants tested, a thousand contained alkaloids of potential importance. This leaves the majority of species in the Plant World—hundreds of thousands—to be investigated.

The quest for plants with important medicinal effects leads to every part of the globe. *Eleutherococcus senticosus*, a Manchurian plant in the ginseng family with no history of use in traditional medicine, was classified and named in 1859. It attracted so little notice that for a century—until Russian researchers screened a number of Far Eastern plants for medicinal activity—the only common name for it was the scientifically incorrect "Siberian ginseng." Today Soviet citizens routinely use extracts of this plant, although Western doctors are largely unfamiliar with it.

According to Stephen Fulder of Chelsea College of London University, the *E. senticosus* compound is restorative rather than curative. Like other substances highly valued in Eurasia and the Far East, where the prevailing systems of medicine emphasize prevention of disease, it may improve the general health and energy of people who are weakened or under stress. Dr. Fulder has coined the term somatensic, keyed to the drug's action in extending human performance. This name literally means "body-expander."

The extract is thought to improve the endurance, the reflexes, and concentration of Soviet athletes. Preparations are also used by Russians in physically active occupations, especially by divers, rescuers, climbers, explorers, soldiers, factory workers, truck drivers, and aviators. Cosmonauts in the Salyut 6/Soyuz 32 space station employed it.

Doctors in Soviet hospitals and clinics prescribe doses of the plant extract along with other medicines for disorders such as anemia, depression, chronic heart disease, pneumonia, and tuberculosis. Soviet doctors also consider it useful for problems of convalescence and old age. The case of *E. senticosus* demonstrates that there are plant drugs used by other cultures that

Below, in the People's Republic of China, specific herbal compounds are formulated from the more than forty-nine hundred medicinal plant species available. Bottom; a postage stamp from Ghana suggests the economic importance of cocoa (Theobroma cacao) to tropical Africa. The bean contains the alkaloid theobromine, a diuretic, and kidney and heart stimulant.

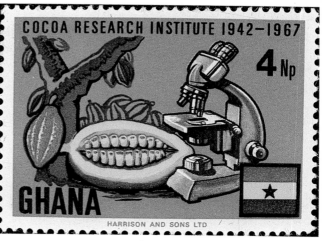

would repay careful study with a view to their possible introduction elsewhere. An outstanding source of research materials is the People's Republic of China where a billion people depend on traditional medicine most of which involves plant drugs. As Norman Farnsworth says:

Can such a medical system have survived for 3,000 years if the entire populace was being served placebo medication?

The same applies to the Hindu system of Ayurvedic medicine in which about fifteen hundred of the drugs used are derived from plants.

Some of modern medicine's most important plants actually face local extinction. These include the crisped glory lily (*Gloriosa superba*), a climbing plant with elongated tubers and beautiful flowers with curved, wavy petals. Their hues vary from cool green to yellow to hot orange-red and they grow in greatest numbers in the coastal areas of India's states of Gujarat, Maharashtra, and Karnataka. The plump roots have been used in the treatment of parasitic skin infections, leprosy, and internal worms, and are valuable in Ayurvedic Medicine, a regional herbalism of long history and considerable renown. The lily's underground, tuber-like rhizomes contain alkaloids, including colchicine—best known for suppressing gout but now tested for anti-cancer properties. Investigators have determined that the seeds rather than the rhizomes are the most economical source of the drug.

Glory lily flowers are so beautiful that masses of them are collected for sale in bouquets, destroying prospects of seed development. It is believed that most of the medicinal material is exported, and collection of seeds and roots for the foreign market causes a shortage of raw material for local drug industries in India.

If wild populations of Indian medicinal plants are allowed to become damaged through excessive collection, a whole series of traditional medicines which have been in use for thousands of years—the Ayurvedic, Unani, GrecoUrvedic, and Indo-Tibetan schools of herbal medicine—will be threatened. Indians are also active in modern biomedical research into the chemical properties of unfamiliar plants. One mint among these species, a close relative of the house plant *Coleus blumei*, is an hypotensive. A member of the hog-plum family may combat cancer.

The old systems of Indian medicine lost official acceptance during the early 1900s as the European system, derived from the canon of the Greek herbalists Hippocrates and Dioscorides, began to supplant earlier doctrines. The uses of a number of plants were thus buried under subsequent "modern" training, and are only now being brought to light and tested. One of these is *Gymnema sylvestre*, a woody climber of the milkweed family from southern India which had been used in the treatment of "honey-urine" for two thousand years. This disease is the familiar diabetes mellitus, which was first identified by

the famous Indian surgeon Sushrutha, who detected sugar in the urine of diabetic victims in the sixth century B.C.

Studies of the hypoglycemic action of dried, powdered leaves of *Gymnema* in diabetes mellitus are now being undertaken at the University of Madras. Given the undesirable side effects that insulin and synthetic hypoglycemic oral drugs can have in diabetic therapy, this research is of great potential importance.

In men, impotence sometimes accompanies diabetes. A new way to study this impairment employs the alkaloid yohimbine, derived from the bark of the African yohimbe tree (*Coryanthe yohimbe*) in the coffee family. It is reputed to be an aphrodisiac. In a study at Queen's University in Ontario, Canada, some impotent patients who received daily doses of synthesized yohimbine achieved erection and orgasm.

The Madagascar periwinkle (*Catharanthus roseus*), a common ground cover, has become a major source of chemicals—most notably vinblastine and vincristine—for the treatment of childhood leukemia and Hodgkin's disease. The therapeutic effect of *Catharanthus* (formerly *Vinca*) *roseus* is mainly derived from alkaloids in its leaves and stems. Their discovery and development represents a remarkable medical breakthrough.

Far left, the bitter gourd or balsam cucumber (Momordica charantia) has edible fruits and leaves, but it is also valued as a medicinal plant in many tropical regions. The leaf is used to treat leprosy, jaundice, and hemorrhoids in Iraq and India, and breast cancer in Nigeria. In India, Sri Lanka, and Iraq, the fruit goes into compounds for rheumatism, gout, spleen and liver diseases, and is also used as a laxative and an abortifacient. Below, during their initiation as herbalists, West African women wear necklaces made of Momordica vines. Dioscorea composita vine, left, is a true yam widely cultivated in Mexico for its yields of the steroid diosgenin—a precursor to cortisone, a remedy for asthma and rheumatoid arthritis, and a component of birth control pills. Indeed, 80 percent of birth control pills produced around the world contain chemicals from Mexican yam tubers.

Strophanthin, a toxic seed glycoside of the West African Strophanthus vine, *above, can cause heart paralysis, but is used judiciously in medicines as a heart stimulant.*

The direct extraction of only one ounce of vincristine requires fifteen tons of periwinkle leaves and costs well over six thousand dollars. Small quantities are so effective, however, that only ten pounds a year are employed in the United States. Eli Lilly markets vincristine by the trade name Oncovin. Vinblastine, known to the trade as Velban, is far more plentiful.

Some nations, including the People's Republic of China, are augmenting the natural supply of periwinkle alkaloids by mass growing of the plants. Because of the low yield of vincristine from *Catharanthus roseus*, the Chinese find it necessary to process up to a quarter of a million pounds of dry leaves at a time in order to recover a reasonable yield of this alkaloid. The leaves represent only about 10 percent of the total dry weight of the plant, so the Chinese usually harvest nearly two million pounds of fresh plants at a time.

Vinca minor, the common periwinkle, has been grown for centuries in gardens. It yields an alkaloid known as vincamine, one currently under clinical investigation in France and else-where. The drug may help to delay senility by increasing the supply of oxygen to brain cells.

In addition to the periwinkles, the dogbane family (Apocynaceae) has given us the valuable snakeroot plant (*Rauwolfia serpentina*). A shrub native to India, snakeroot has been used here medically since 2000 B.C. The root itself yields more than 75 alkaloids, including reserpine—first isolated in the 1950s. Important in sedatives and tranquilizers, reserpine is also used in the treatment of such mental derangements as paranoia and schizophrenia. It also effectively alleviates hypertension and complications of high blood pressure and strokes, heart disease, and kidney failure.

Because laboratory synthesis of reserpine costs $1.25 per gram, plant material is more commonly used for commercial preparation, at the cost of approximately 75 cents per gram of the purified alkaloid. Due to reliance on natural sources for the commercial supply, this plant became threatened by overcollecting. To conserve it the government of India has placed an embargo on the export of the roots.

In some instances, market demand can threaten the survival of these species in the wild. Yet, as with the periwinkle in China, propagation of medicinal plants in the developing world can reduce the erosion of genetic resources and create business opportunities.

There is currently an enormous surge of interest in the use, development, and conservation of medicinal plants throughout the world. Several official bodies are involved, including the World Health Organization (WHO) whose effort to integrate modern and traditional medicine in the developing world is outlined in the next chapter. The United Nations Industrial Development Organization seeks the establishment of local pharmaceutical industries based on wild plants, and the OECD is planning a major program on the economic value of medicinal plants. Traditional medicine also has its enthusiasts in the industrial nations, where herbal teas are so popular that some of their plant sources are locally endangered.

Above, a traditional medical practitioner in Ghana wields a fly whisk over a bowl of herbs used to purify a shrine for the graduation ceremony of herbalist trainees. Left, the open pod of Griffonia simplicifolia reveals two discoid seeds attached to the suture. Various parts of the plant are antiseptic. The bark is used in plasters for chancres, the leaf juice as an enema and for kidney disease remedies, and extracts in compounds used to treat vomiting, diarrhea, and constipation. Reputed to be an aphrodisiac in its native West Africa, the legume contains a chemical which exhibits characteristics similar to those of LSD.

Visit to the Doctor

A neatly dressed mother and son wait with others who have come to a village near Nairobi, the capital of Kenya, to be treated by a renowned practitioner of traditional medicine. In East Africa, as in most places on earth, Western scientific medicine has been generally available only since 1900. In fact, medical treatments were largely "traditional" even in Europe and North America until scientific medicine became widely established after the 1920s.

As a young woman living under colonial rule in Kenya, Elizabeth Karu practiced what Europeans called witch-doctoring. Exorcism was her specialty, and even after it was outlawed she continued to offer help in secret. Now Kenya's government fully supports her healing mission. Though exorcism is alluded to in Christian tradition as having been practiced by Jesus Christ, belief in it went out of vogue in much of Europe and America with the spread of the scientific method.

THIS YOUNG WIFE AND HER ailing son appear frightened but undaunted: common enough in any doctor's waiting room. These two would not have come unless they believed in, or wanted to believe in, the power of the healer. He is Barnabas Kiriu, a renowned herbal physician of 50 years experience. Jomo Kenyatta, Kenya's first president, consulted him.

The boy probably suffers from sickle-cell anemia. It is one of the many afflictions which Barnabas Kiriu encounters in his practice. Hundreds of people come each day from the Nairobi area and more distant parts of the country to be received by the famed practitioner and his associates.

Before seeing the herbalist, though, patients must pass close to Elizabeth Karu, a person likely to be termed an exorcist by westerners, but who is called a witch doctor by her own people. Her elaborate ritual involves special music played on a concertina and the casting of shells for clues to the origin of disease in some of her patients. Elizabeth Karu cleanses the spirits of those who come for therapy to their bodies.

After deciding on a course of treatment, Barnabas Kiriu directs his team of herbal "pharmacists" in the preparation of specific tonics and other compounds from woods, bark, and other biological materials. Increasing numbers of traditional practitioners now refer acute infections and medical emergencies to hospitals and clinics. On the other hand, scientific medicine is sometimes at a loss to cure conditions which herbalists treat successfully.

Once Elizabeth Karu has questioned patients, seeking clues to emotional stress that can trigger physical symptoms, groups of a dozen or more patients are treated by Barnabas Kiriu, a specialist in herbal therapy. Having spent half a century seeking cures locked up within local plants, he has devised treatments which, like those of Elizabeth Karu, function within a cultural setting. For example, his "group therapy" is designed to reassure Africans, many of whom dread isolation, even in a doctor's examination room. The laying on of hands, above, further relaxes the sick persons, who also seek support from family members and friends who often accompany them. In this manner, Mr. Kiriu works with as many as two hundred people a day.

Above, herbal elixirs are bottled for distribution to patients. Barnabas Kiriu explains dosage to the concerned woman and her ailing son, left. Analysis of medically active plants used in the compounds of village healers can prove beneficial to mankind—as in the case of the cure for childhood leukemia presented in the last chapter.

The WHO and Herbalists

Traditional healers of many lands recognize that physical ills can have a psychological background, and that the spirit can help to cure the body. Working with this knowledge, and with herbal medicines, the herbalists of the developing world manage to care reasonably well for those three persons out of every four in our world who consult them, either out of necessity or choice.

Furthermore, with greater understanding between old-style healers and modern physicians, most of the people of the world can profit from improved health care before the end of the century.

Indeed, western doctors already depend heavily on medicines derived from plants, as we have seen in the preceding chapter. In addition herbal healers, particularly of the tropical rain forests, can lead scientists to new cures—just as their use of Madagascar periwinkle in treating other diseases lead to its development for cancer chemotherapy by western scientists.

With such facts in mind, officials of the World Health Organization (WHO) have initiated a program designed to provide improved health care for the developing nations by A.D. 2000. Toward this end, encouragement is given to governments in Africa, Asia, and the Americas to assist in the training of village herbalists. These traditional practitioners can improve their own skills and at the same time learn to recognize diseases for which better cures exist in scientific medicine. Such a dual health delivery system has arisen in the People's Republic of China during the post-war period. Members of the WHO work to bring about greater cooperation in medical matters between the industrialized nations and those which seek greater economic and social development.

PART IV

A Paradise To Save

DURING THE LATE 1970S AND EARLY EIGHTIES deforested hillsides like these in sight of the Himalayas have become symbolic of destructive forces at work throughout the developing world. Yet in a few hidden folds of hills and mountains, climax forests survive. As here in Nepal, local reforestation has begun. There are already indications that the economic advantages of tree farming for fuel and other uses are beginning to be recognized throughout the developing nations. Greater value is being placed on preserving forests, as in Lebanon where the majestic Cedars of Lebanon are generally affirmed as sacred by all political factions of this war-torn land. Early large-scale hardwood plantings are proving their worth, notably at Addis Ababa, Ethiopia's capital, and on hillsides in the Republic of Korea. In the People's Republic of China, many of Beijing's streets are lined with trees, as with many other cities and towns throughout the world. But here careful pruning of urban forests yields important amounts of fuel. Bringing production into better balance with harvest is surely one of the most important global issues of our day.

In the four chapters of this section, the authors and the institutions they represent urge scientists, business people, and other citizens to consider and promote conservation at all levels from local to global. The first two chapters deal with the anatomy of global problems. The final chapters provide specific examples of materials and techniques for helping to save the green world and for promoting the well-being of the majority of the world's people—the people who live closest to the land. Many of today's scientists are recognizing that village farmers, woodsmen, foresters, and other people associated with rural areas and wilderness can undertake best the work of *conservation for production*, a central tenet of the World Conservation Strategy and of the national conservation strategies noted throughout this book.

Enduring Impact

IF THE MIND COULD CREATE A TIME-LAPSE film show of ecological history, we would see that green nature—the so-called natural landscape—is in a sense largely artificial in most inhabited areas of the world. Today we are cutting through second or even third growth of forests where land had once been laboriously cleared and cleared again for farming during waves of settlement in centuries past and during recent generations. As we have seen, even the delightful landscapes of the Mediterranean are mostly man-made.

We do, however, possess a way to help us see the past unfold. By means of space satellites, we can peer down at immense areas of the globe and take pictures of what we see. Properly interpreted, the images reveal patterns of change, as with new urban growth branching off from old, or jungle damaged by fire. In the case of the great vista of Bangladesh at the Ganges-Brahmaputra delta, opposite, green has been transformed electronically into red for clarity. Each dot likely signifies a single farmstead. The intensity of human activity, and thus of ecological impact upon the land is immediately clear.

In a sense, sophisticated interpretation and evaluation of high-altitude imagery does let us also look into the future. To begin with, we understand that when natural disasters occur, the amount and the kind of human tragedy that results is a function of the density of human population and the ways in which the land is used. Of course, more people mean additional casualties, but a multiplier-effect can also come into play. For instance, floods in most parts of the world are thought of as natural catastrophes, and in one sense they are. More and more often, however, human activity lies behind an almost exponentially rising scale of disaster. Examination of a common situation in many tropical lands shows why.

Rivers were once the only highways for many villagers. Simple networks of trails fanned into the interior from settlements on the shores—the blueprint for development. Roads were built to supplement and replace the tracks and river routes. Villages grew into towns and the movement of resources followed the roads. Ever more intensive agricultural development was required to feed ever greater numbers of people. Forests were leveled to provide fuel and the undergrowth stripped or burned away to open up more land for agriculture.

Once the land has been so altered, destruc-

Bangladesh, one of the most densely populated nations on earth, occupies much of the great river delta of the Ganges and Brahmaputra at the Bay of Bengal. Color applied to aid analysis of this 1977 Landsat image—made in February, the dry season, when many crops are ripe—generally represents plant life in red. Continuous areas of red are the contiguous fields of individual farms; most fields shown are probably rice. In some areas, individual farms are shown as tiny red dots. Such Landsat images are as useful as aerial photographic surveys for analysis of river meanderings and land use and at only one percent of the cost. S. Dillon Ripley, Secretary of the Smithsonian Institution, was an early advocate of overhead imagery for the assessment of the natural resources of the Indian subcontinent and neighboring areas.

tion can occur on a grander scale than before. Eventually, rains scour the land. Swollen with the swift run-off of rainwater, streams quickly cut deep ravines. With the retentive plant cover gone, little remains to divert the flow of water on its way to the streams, to soften its erosive power, or to slow down the life-giving fluid long enough for it to soak into the earth. Unleashed, the waters may surge into large waves that race across the plains. People and cattle perish. Standing water destroys crops. The land is slow to recover and receptive to the recurrence of dis-

aster on a rising scale.

Such tragedy unfolds in several different parts of the world including Bangladesh. Disaster has roots upriver in Nepal, where the hillsides have been laid bare with the harvesting of wood for fuel. As monsoon rains saturate the hillsides and parts of entire villages slide down mountains, millions of cubic yards of Nepal wash away toward the Ganges-Brahmaputra delta at the Bay of Bengal.

Despite the damage to the national legacy, Nepalese peasants must live; then, for the present, they must cut the remaining trees that hold the soil. Here as elsewhere, the only resource that most people of the developing world possess is the land on which plants already grow. There is little mass of vegetation in semi-arid regions compared with the huge rain forests. In both instances, people must use what they have. The problem is how? Rapid population growth complicates the situation.

Imagine that you live in a semi-desert region and your wealth is measured by the number of goats you own. You wish, of course, to see the herd steadily multiply. You also want your children to possess wealth of their own, so the number of herds also increases. Under this pressure—the stress of teeth and hoofs—the edible vegetation of the wet years is steadily diminished. Plants are unable to produce flowers and seed-bearing fruit. Seeds already in the sand—nature's savings bank—keep germinating with each rain until they are exhausted.

Then come the dry years. Your animals and those of your children die off. When the rains finally arrive, the stumps of the edible plants are so severely exhausted that only a little regeneration occurs. Few seeds remain in the soil to sprout. So steadily the deserts grow.

Threats to the countryside from agrarian life are more than matched by those from urban build-up. In many parts of the world, in both developed and developing countries, people are moving away from the land to inhabit cities and towns. In the industrial nations, many central urban areas are suffering decline. Urban exodus leads to ring upon ring of sprawling suburbs. Here, too, the effect is destructive, creating no new resources and increasing the pressure on the remaining countryside and agricultural land. This situation, already dangerous, can soon turn tragic when fueled by the world's explosive reproduction rate. While projections differ, at least a fifty percent rise in the world's population is expected between the year 1980 and the beginning of the next century.

Settlement widens; people cut deeper into nature's green. Now that satellites constantly survey the land, we possess a detailed and accurate account of the events that alter the surface of the earth. The practical application of this knowledge is much more advanced than many people realize. India, Nepal, Bangladesh, and Thailand, for example, have national centers for the analysis of Landsat and other imagery from space.

Fire, soil erosion, and animal grazing all place destructive stress on the land, progressively diminishing the productive value of farm, forest, and pasture acreage. Opposite; fire engulfs sections of the Ivory Coast's Tai National Forest, a woodland poised between the semi-arid Sahel and the more verdant coast. Rapid erosion after land clearance in Mexico, below, presents a sobering study in green and ochre. Bottom; goats and other grazing animals—their numbers rising—threaten plants and soil near Saint-Louis in Senegal.

A fence surrounding New Zealand's Mt. Egmont protects forests on the volcano's slopes. The areas outside the enclosure, like the New Zealand site below, show the effects of deforestation and grazing.

Knowledge derived from the pictures can be applied to problems of resource management, demographics, agriculture, fishing, and to relief and rescue situations in times of natural disaster. Since 95 percent of all floodwaters come to Bangladesh from outside the country, this populous nation has established a flood and storm warning facility at Dacca. With conditions constantly monitored, emergency aid can be prepared in advance and delivered where needed most. With its active space program, India receives picture transmissions directly from Landsat, as does Thailand. The means are available to manage our resources but will these fine examples be followed worldwide? And is there time and sufficient willpower?

Forest in Viet Nam lost its leaves after being sprayed with defoliants during military operations. The trees remain and will recover.

206

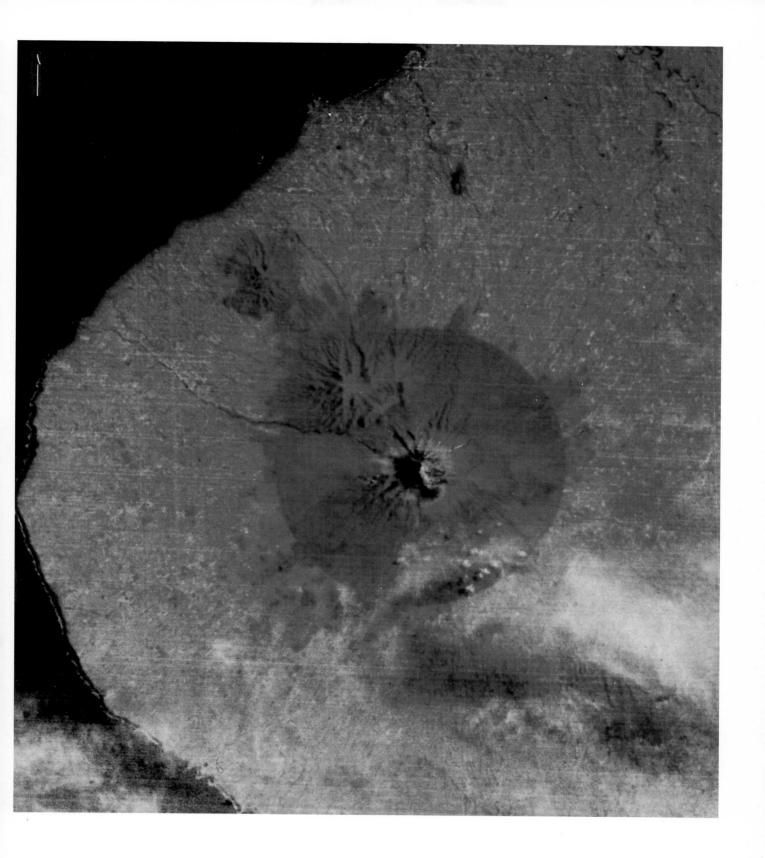

207

Genetic Erosion

THE PRECARIOUSLY NARROW GENETIC BASIS of many of our crop plants may come as a surprise. For example, four varieties of wheat produce 75 percent of the crop grown on the Canadian prairies. More than half of the prairie is sown with a single variety. Four cultivars of potatoes account for 72 percent of production in the United States. All of the coffee trees in Brazil have been derived from the seedlings of a single plant cultivated in the Amsterdam Botanic Garden in 1709. The entire United States soybean stock came from six plants from one place in Asia.

The consequences of relying on such a small range of genetic material for our crops can be disastrous. Sri Lanka (formerly Ceylon), for example, was a major coffee producer until the crop was decimated by disease in 1869. Due to an insect pest, phylloxera, virtually all of Europe's grape vines were destroyed between 1870

Wheat in vast North American fields yields enormous crops, but in some places cultivation or other use of the land may destroy wild species whose genes could one day save crop species from disease.

and 1900. Since rootstocks from wild grapes were resistant, they were grafted to European stems and the vineyards recovered.

Likewise the cultivation of genetically uniform, high-yielding, and advanced cultivars carries with it very high risks of total crop failure through disease or pests. It is for this reason that there is a drive worldwide to conserve genetic resources of crop plants and their wild relatives so that there are genetic reservoirs available to the plant breeders to provide the raw material for future adaptations and for tomorrow's new cultivars.

Today, all kinds of genetic permutations exist—some expressed in overt traits, but others hidden in the genetic code. One by one we discover them as we classify new species, study their evolution, or hybridize new plants for agriculture or ornament. Botanists and breeders must therefore tread their paths amidst a bewildering array, in highly complex species, of offbeat traits described by such technical terms as apomicts, hybrid swarms, clines, clones, lines, strains, mutations, cytotypes, races, varieties, and cultivars.

Due to the extraordinary variability found within plant species, there are large numbers of uninvestigated alleles—genetic traits—in many of them, and it is difficult to overestimate the importance of conserving such genetic potential and diversity. The wild relatives of domesticated plants can often provide genetic "refreshment" when current varieties become susceptible to disease or are otherwise threatened. These are also used in breeding for improved yield and nutrient content, as well as in the production of new strains of crops. The chromosomal and genic variation within plants might be called a magician's garden from which new marvels can appear at any time. Genetic material introduced from wild species has already helped to double the yield of sugar cane. Yields of wheat, rice, and other grain cereals have risen dramatically around the world through newly created strains—the so-called Green Revolution.

A major problem with today's methods of large-scale farming is that the highly-bred, uniform strains of crop plants are grown over such vast areas that one epidemic could wipe out the whole season's harvest. The average lifetime of wheat cultivars in Europe and North America is only five to fifteen years because pests and pathogens themselves evolve new races that combat the genetic resistance of plants.

This means that there must always be an effort underway to introduce new vigor into many crops. In many cases wild relatives are the only ones with inborn resistance. Therefore, it becomes imperative to preserve as many of these primitive cultigens, or land-races, as possible, seeking them out in the habitats where they have been discarded in favor of advanced cultivars that are vulnerable to disease over the long run. Since conservation *in situ* is seldom feasible for cultivated plants, the genetic resources have to be stored in seed banks, germ plasm collections, or by tissue or meristem culture. The loss of germ plasm of the wild relatives of crop

Physical erosion occurring beside this highway in Brazil can cause genetic erosion as increasing areas of virgin rain forests are briefly cleared, then exploited, and finally abandoned when the fertility of the soil has been exhausted.

The global market for orchids, long prized by florists and hobbyists for their exotic beauty, has brought many species to the edge of extinction. Demand for the white nun orchid, Lycaste skinneri, *was created through advertisements such as the one from the late 1800s, below. Although it is native to montane cloud forests throughout Central America, local collectors and destruction of its habitat have threatened this most commercially important* Lycaste *species, and it is now on Appendix I in the CITES treaty. Similar forces have also made* Drymodes picta, *right, exceedingly rare in its native Burmese forests. Hired by the famous naturalist Karl von Martius, a local botanical collector strips an orchid-festooned tree near coastal Ubatuba in São Paulo province, Brazil, ca. 1817.*

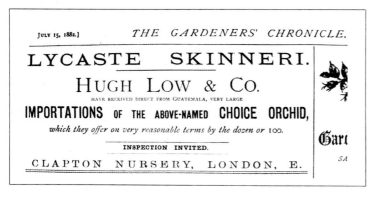

plants—genetic erosion—may be caused by loss of habitat leading to the contraction or even extinction of populations or to overcollecting, sometimes at the hands of specialists seeking the exotic, the rare specimen, the bizarre.

The horticultural industry sometimes acts as a valuable force in conservation by the propagation and distribution of rare species. This trade, as in the case of agriculture and forestry, has to replenish its stocks from the wild. In the case of bulbs, rock plants, cacti and succulents and other special groups, this can lead to exploitation of wild populations by rogue traders who cater to the demands of unscrupulous or naive gardeners. Often the plants are badly packed and do not survive shipment. As can be imag-

ined, such rogue traders are not popular with the legitimate trade. They give the industry a bad name.

Then there are specialist horticulturalists, both professional and amateur, who concentrate on particular groups such as orchids, ferns, succulents, and alpines. They have a very important role to play. They may successfully cultivate a very rare species and make it common; thus helping to remove pressure from the wild population, while fulfilling the needs of the trade and helping with introduction of rare species back into the wild. Sadly, they too have their "traders" as we shall call those people who rob the wild continuously to satisfy their clients' needs. They often focus their attention on the rarest species—for these fetch the highest prices. There is a particular type of person who craves to own the rarest, preferably the only one. When the object coveted is a plant, this sickness is a real threat to many species.

Cacti and succulents are often subjected to tremendous pressures through collecting in the wild, as in North America and southern Africa. Fortunately, techniques are being worked out to propagate and cultivate plants and thus fulfill the demands of collectors. In 1974, the International Organisation for Succulent Plant Study, better known as IOS, prepared a code of conduct for its members, a splendid example of ecological awareness that can be valuable to all horticultural societies and plant enthusiasts.

Similarly, the cycads—those palm-like plants that have persisted with no real evolutionary change for more than 50 million years—are now collected too heavily. It is sad to think that the greed of people today might finally drive some of these noble species into extinction. For example, in her book *Cycads of South Africa*, Cynthia Giddy points out that in the Transvaal of southern Africa the cycad loss from the wild was so rapid that protective laws were enacted in 1971. To control the trade, all owners were required to have their plants licensed. Eight thousand permits were issued to owners of sixty-four thousand plants during the first eighteen months of the program.

Even the harvesting of cut flowers may be a threat to native species as in Western Australia, an area rich in endemic species. Since many of them set few seed, commercial collecting can lead to serious dangers of extinction.

In some cases, though, heavy trading can be advantageous. There are always situations where a particular region is to be razed or flooded for economic reasons; collecting and selling plants here is thoroughly justified. Conservation plans should also allow for setting up a local horticultural trade. This is best designed to give a steady production and income at the local level, rather than a quick selling off of the wild stocks.

Even botanists can be serious offenders when collecting specimens of rare species for preservation in their private collections of dried plants. This practice is much less common than in the past but it still exists. Sometimes this is encouraged officially by institutions which col-

Far left: Vibrant bloom of Renanthera imschootiana, *an ornamental epiphytic vandoid orchid being rapidly depleted from the forests of Assam, Laos, and Viet Nam. At left, fertile fronds of the curly-grass fern,* Schizaea pusilla, *which inhabits bogs of New Jersey subject to drainage and development. Opposite; a tableau of nine endangered, threatened and vulnerable species. Top row, left to right:* Cypripedium calceolus *(Eurasia),* Orchis simia *(Great Britain),* Tulipa greigii *(Turkestan); Middle row, left to right:* Orothamnus zeyheri *(South Africa),* Primula palinuri *(Italy),* Vanda coerulea *(India); Bottom row, left to right:* Richea scoparia *(Tasmania),* Paeonia cambessedesii *(Majorca),* Tulipa kaufmanniana *(Turkestan).*

lect herbarium specimens in sets of a hundred, sometimes including uncommon species, for distribution to other herbaria.

In developed and developing nations, technological change can bring unexpected genetic erosion. An example from Britain shows dramatically one small effect. Britain is broken up into one hundred and twelve botanical sub-divisions or vice-counties. The Corn Cockle (*Agrostemma githago*) was once widespread in grain fields occurring in a hundred and four of the vice-counties, many local floras recording it as "common." The plant began its disappearance with the improved seed-cleaning techniques introduced in the 1920s. With the advent of herbicides in the 1960s, its decline became even more rapid. Extremely rare today, it is found in only two counties in England and one in Scotland. Its only hope is the gardener, and the garden—the court of last resort for saving this beautiful flower, one of considerable horticultural merit. If the Corn Cockle and its many companion cornfield weeds can be maintained by horticulture or appropriate "subsidized farming methods," so much the better.

Many other deliberate changes have their

effect. Selective removal of plants poisonous to cattle can drive species to extinction. Yet might not those very toxins in the plants have a value to mankind, perhaps in a medical context? We might profit from maintaining poisonous plants in reserves while their value is assessed, for we possess the tools for complex biochemical assay, but not the time. Little by little, and we hardly ever realize it until too late, our options are growing smaller and smaller.

We are pleased to note that recent advances in cancer chemotherapy, based on alkaloids, have begun to revive interest in mass screening of plant tissues for medical activity. Edward Ayensu has predicted that researchers may very well find a plant source of interferon, or a chemical precursor of this material which can fight viral infections and may have anti-cancer properties. Most such superdrugs probably are to be found in the equatorial regions.

In the tropics, technical innovation can damage the genetic base. Introduced species are often used to establish plantations and agribusiness methods are also imported. Through such onslaught, many native animals, insects, and other plants are lost—with damage to the whole

web of vital ecological relationships.

Drainage of wetlands for farms can mean the loss of native species. Here, too, simple changes in agricultural machinery and the handling of crops prevent arable weeds from being broadcast with crop seed and perpetuated. This is good for agriculture and thus good for us—in the short term—though the loss of these species may be a cause for concern and it may well be that some of them should be maintained in special reserves.

As we have seen in this chapter, genetic erosion threatens a whole series of different kinds of species, even relatively common ones. This erosion means that some of the original pool of genetic variability that existed in the species has been reduced through the destruction or loss of large parts of their distribution area.

Professor Jack Hawkes, a leading expert on the conservation of genetic resources, has summarized the position:

As forests are felled, marshes are drained, sea coasts are turned into holiday resorts, mountain pastures are trampled and grazed, heathlands are changed into grassland, old rich meadows are ploughed up and planted to crops or re-seeded with standard grass mixtures, cattle and sheep grazing is intensified, cities expand, industry spreads and roads are widened, so the genetic diversity of all plants in every part of the world is diminished.

The widespread damage that the green world is sustaining, through physical and chemical damage to the land and waters, and genetic erosion is far more serious than most people realize, although the cumulative effects leading to an environmental tragedy might take five or ten decades to unfold.

Logging of trees causes loss of associated forest cover species, evidenced by the forlorn cutovers in the Lonquimay area of Chile, at left, and the eroded wasteland of Philip Island, at right. Above; Streblorrhiza speciosa, *the extinct Philip Island glory pea, was endemic to the island.*

Underexploited Crops

Food plants are closely linked with various cultures. The broad bean of Egypt, opposite, has been cultivated continuously for four thousand years. In tragic contrast, the grain amaranth, below, died out in pre-Columbian Mexico when the Spanish crushed the Aztec civilization beginning in 1519, but may once again prove valuable to the region, and spread to other areas.

TWENTY-SIX INDIVIDUAL AMINO ACIDS, THE chemicals that make up protein, are required for the biological functioning of the human organism. Each day, each human needs enough of the right kinds of proteins to assure proper growth and repair of living tissues, including those of the brain and nerves. Ample amounts of particular amino acids are vital for the unborn, infants, and children. Unfortunately, reliable supplies of protein and other food substances cannot be taken for granted in our world.

An important study by the United States National Academy of Sciences, *Underexploited Tropical Plants With Promising Economic Value*, points to many cereals, roots, tubers, vegetables and fruits which deserve wider attention by cultivators in the world's tropical countries, where much of the world's hunger is found.

Among underexploited cereals are grain amaranths (*Amaranthus*), which belong to the same family as the showy garden plants cockscomb (*Celosia*) and the Jacob's coat (*Alternanthera*). They are very high in lysine, an essential amino acid generally scarce in vegetable protein. Amaranth numbered among the staple crops in the highlands of Mexico, Guatemala and Peru during the early sixteenth century. Spanish conquistadors suppressed cultivation of this highly nutritious grain by the Indians and introduced barley, a cultivated grass lower than amaranth in some important amino acids. Amaranths also have edible leaves which are used as a spinach in Nigeria and other African countries, in the Caribbean, in Greece, and elsewhere.

Along with tropical grasses such as sugar cane, amaranths are among the special plants referred to as C4. They perform photosynthesis at a higher rate than most other plants and can also germinate and grow in a wide range of climates and soil types. Corn, another native of Mexico or nearby areas, also benefits from C4 vigor. Its immense agricultural usefulness around the world could probably never be matched by amaranth, or by most other underexploited plants, but this "Golden Grain of the Aztec" could become very important. To achieve widespread success, unfamiliar crops usually need their champions—researchers, plant breeders, and publicists.

Buffalo gourd (*Cucurbita foetidissima*), one of many wild species belonging to the squash family, the Cucurbitaceae, has been used by American Indians for centuries. The plant, which occurs from South Dakota to Mexico, is

The cucurbit family (Cucurbitaceae) comprises 90 genera of climbing and trailing plants native to temperate and tropical regions of the globe. The more than seven hundred and fifty species have unisexual flowers and, in many cases, a type of fleshy fruit with tough rind and numerous seeds that has been given its own botanical name: pepo. Familiar and economically significant pepos include pumpkin, cucumber, squash, watermelon, cantaloupe, the loofah sponge, and a myriad of useful and ornamental gourds. Many cucurbits contain toxic chemicals which have been used in purgatives and emetics. The edible buffalo gourd (Cucurbita foetidissima), a wild roadside plant of the western United States and Mexico, has recently undergone breeding trials for domestication. Clockwise from right are its pulpy interior, its large subterranean root, the young fruit and its edible seeds, and the flower visited by a pollinator.

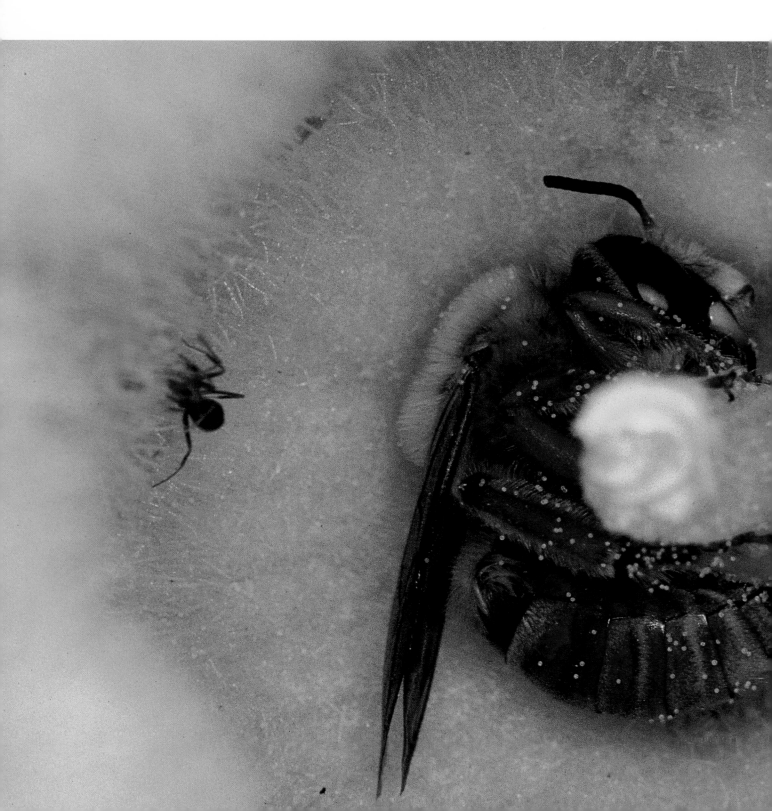

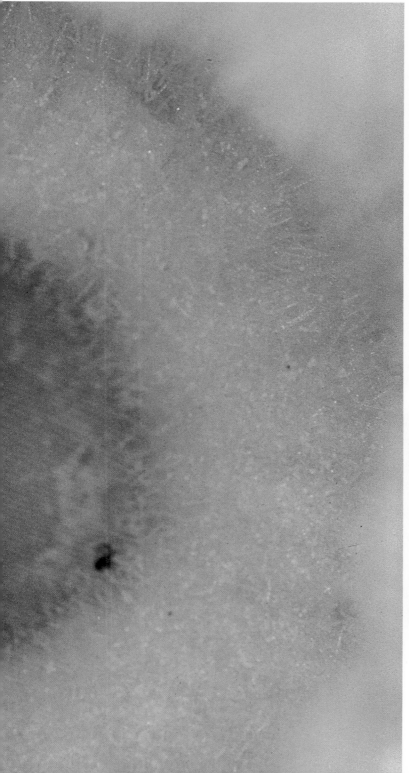

being transformed into a commercial product.

The years of work could bring an important economic advantage. Buffalo gourd grows vigorously in places inhospitable to traditional cultivation—soils only marginally suited to agriculture—such as dry wastelands, stony prairies, and a variety of semi-arid areas. The fleshy stor-

At right, a greengrocer in Bangkok sells tropical winged beans. Virtually all parts, including the tuberous roots, the pod, above, and the seeds, top, can be eaten.

age root penetrates as deep as fifteen feet, attaining a weight of 60 pounds, half of it starch. Each of the perennial ground vines develops 60 or more yellow gourds up to three inches in diameter. By weight, each of the three hundred white seeds embedded in the pulp is nearly a third nutritional protein and 40 percent edible oil. This compares favorably with peanuts and soybeans, and an acre of ground can produce a ton of seed.

The plants themselves are propagated by cloning from the prolific roots which sprout from nodes on the runners. Plant breeders, however, usually employ sexual reproduction to hy-

bridize varieties for early maturation, increased seed yield, and other advantageous factors.

The palm family also offers several potential foods which could be more widely grown and harvested if suitable attention were given to their cultivation. The fruit of the peach palm or pejibaye (*Guilielma gasipaes*) has a good balance of protein, oil, carbohydrate, and minerals. It grows widely in Central America and northern South America.

Each peach palm sends up several independent trunks, which makes it a good candidate for the production of hearts of palm, also known as palm cabbage (although it tastes more like artichoke to some people). The growing tip of the palm tree is a cylindrical bundle of leaves and meristematic tissue that cannot be replaced by the plant when harvested. Thousands of trees have been used in this way for local consumption in Latin America and the West Indies, and some palm hearts are canned as a luxury food for export. Species of several kinds of palms are recommended for growing on a plantation scale; *Euterpe oleracea*, currently cultivated in Brazil and Costa Rica, can be harvested in three years.

The bean family, Leguminosae, offers much hope for new food with value as protein supplements. Pulses, or edible legume seeds, are probably second in importance only to cereals for human and animal food. Some contain two to three times more protein than cereal grains which are cultivated grasses. Known from Biblical times and reported in the Old Testament Book of Daniel, the exceptional food value of pulses can be exploited globally in our times.

Many early people discovered that beans and other legumes provided an especially wholesome diet when eaten with a little grain or bread. Inspired cooks from many lands through the ages have invented such traditional favorites as hoppin-john, red-and-white, succotash, turtle beans and rice, Oriental fried rice with bean sprouts, and Egypt's spicy sausage cake of deep-fried fava beans tucked into the pocket of a wheaten pita loaf.

Many legumes are as valuable for their foliage as for their seeds. They also enrich the soil through ammonia-producing bacteria that live in nodules on their roots. And since the range of these plants is global, domestic varieties can be drawn from genetic stock adapted to particular climates, soils, and other geographical factors.

Tarwi (*Lupinus mutabilis*) has been grown by the Indians of the Andes for fifteen hundred years. This pre-Incan crop is still cultivated in Ecuador, Bolivia, and Chile. It is adapted to cool tropical highlands as well as temperate regions. Up to half of the seed's weight is protein, though they must be soaked in running water to remove the bitterness caused by an alkaloid. Some sweet ones have now been bred, and this development could make tarwi a major crop. The National Institute of Nutrition in Lima has undertaken a program to make tarwi a part of the everyday diet of Peruvians. The value of this crop is being publicized through the distribution of recipes featuring tarwi in combination with corn and other foods.

Rice bean (*Phaseolus angularis*), with its protein content of 20 percent or more, prefers the humid lowland and is moderately resistant to drought. Native to South and East Asia, it was once widely planted after rice had been harvested—hence the name. Today's farmers, however, often harvest several stands of rice a year, one after another, in a practice called multiple cropping. This approach is common, thus there is no chance to plant the traditional rice beans. Such a decline is unfortunate in terms of social and cultural continuity. It is also nutritionally unwise. A great initiative of the 1960s and 70s, the Green Revolution emphasized the cultivation of new strains of wheat and rice. Today we are beginning to augment the foundation grains with other crops, including pulses. Since there are so many legumes, varieties can be found that will flourish under most conditions. Production of more kinds of crop species can help complete the work begun by such innovative economic botanists as N.E. Borlaug, winner of the 1970 Nobel Peace Prize.

As a promising legume for the tropics, the winged bean (*Psophocarpus tetragonolobus*) has few equals, and except for stems, every part is edible. Until recently, this vine was grown almost exclusively in Papua New Guinea and Southeast Asia. Its remarkable attributes went all but unnoticed by scientists, although much appreciated by local people. Today cultivation of the winged bean has begun in Western Samoa, Hawaii, the Philippines, Sri Lanka, Ivory Coast, and Ghana.

This plant has been called a supermarket on a stalk. Among its merits are seeds with up to 37 percent protein—same as the soybean—and pods that are eaten as a fresh vegetable. These resemble green beans, but are often larger. The leaves are eaten like spinach. Steamed, the flowers

Legumes are not all low meadow plants like alfalfa and the clovers. Trees such as *Leucaena leucocephala* at left, which grow to 30 feet in just one year, have begun to provide firewood, forage, and soil improvement in the tropics. Pulse, the edible seeds of legume pods, below, contains proteins that complement those of grains and thus can provide balanced nutrition to various historically malnourished tropical populations. Asiatic *Parkia* is displayed.

Manatees, also known as seacows, are voracious vegetarians; their feeding habits can stem the onslaughts of water hyacinths and other invasive aquatic weeds in tropical waterways.

taste like mushrooms. And below the ground the firm-fleshed tubers have five times more protein than potatoes.

Plants native to the tropics may also provide fodder. One candidate, *Acacia albida*, grows from southern Africa to the tropics and into Cyprus, Israel, and Lebanon. It has the ability to bear its leaves in the dry season when other plants have lost theirs: thus its value is heightened. Not only do cattle, goats, sheep and camels relish its leaves, wet or dried, but during the wet season when *Acacia albida* is without its leaves, other crops are grown in the groves.

Also underexploited, a tree of the mulberry family can provide multiple benefits. When weather parches other plants, *Brosimum alicastrum* supplies reliable fodder. Native to lands near the Gulf of Mexico, it is already a major livestock feed in the Yucatán, Guatemala, and Belize during the dry season. Juicy leaves and branch tips go to the cattle, and pigs eat the fruit. People like the seeds and a white latex which flows from tapped trees. For this "milk" it is called the cow tree. In its native habitat, the plant is considered a prime source of feed—the equal of grass from the best pastures.

The best cattle food plant for the seasonally dry tropics may well be *Leucaena leucocephala*, a leguminous shrub or tree. Native to the midlands of southern Mexico, Guatemala, El Salvador and Honduras, leucaena has been spread to lowland areas of the Pacific and Caribbean coasts of Central America. A variant known as the Peru type grows to fifty feet with extensive

224

One water hyacinth portends a million. The Sri Lankan at left is fighting a losing battle at Lake Beira with these plants, among the fastest growing in the world. The scene is repeated from Bangkok to Brazil. Stately violet "hyacinth" flowers are so attractive that hapless people eagerly carry the plant back to their homelands—and often live to regret it. Nonetheless, uses may yet be found for this invasive aquatic weed, which has been ground up for mulch and animal fodder. The newest application is at Disney's Epcot Center in Florida, where the water hyacinth's dangling roots absorb pollutants from waste water, making it an effective living filter.

branching, but is not allowed to grow tall when its lower branches are used as fodder.

Leucaena grows best in limestone soils of tropical lowlands where it can survive long dry seasons. Here, as elsewhere, it provides firewood. It coppices easily. That is, leucaena withstands severe pruning and quickly regenerates the missing parts. In fact, on an overall basis, leucaena may be more valuable for its wood than for its edible parts. Some species grow remarkably fast.

One drawback has been observed in Australia where cattle fed extensively on leucaena lose their hair and suffer infertility and other adverse effects. An amino acid known as mimosine is implicated. So here leucaena is often mixed with grass or other food. Strains with

little or no mimosine content may soon reach the market. This disadvantage can become an advantage at shearing time. After sheep have been fed on leucaena, their wool can be removed by hand, simply lifted off the animal's back.

The wealth of plants mentioned in this chapter is only a sample of the many species which provide new opportunities to feed and sustain the starving millions and improve the quality of life for millions more.

Conservation for Production

island of Hispaniola in 1492. Since gold was found soon after, colonial exploitation began early. The native Arawak Indians were enslaved and soon died out. The land suffered, too, and five centuries of harsh treatment have denuded many hillsides, especially in Haiti, the western part of the island and the hemisphere's most populous island nation. Soil has eroded, adding to a long list of economic and environmental problems. Reforestation, as shown opposite, has begun not only to provide cover for the ground but also to help prepare the way for more ambitious projects in years to come. Integrated development plans with full environmental overview are proving their worth every day.

Ever since ecological problems emerged after the creation of Lake Nasser behind the Aswān High Dam, the World Bank and other development agencies have paid increasing attention to environmental factors. What good does it do, for instance, if upstream vegetation is so damaged by peasant agriculture that the soil, and therefore the life of the land, ends up washed into a reservoir behind a dam built for hydroelectric generation and flood control?

In India and other lands, plantings protect the land and thus prevent erosion. The vegetation also yields fuel or fodder. A closer alliance of conservation and production presents a doubly effective attack against poverty. It also reflects growing sophistication by both national and international developers, and represents a strategy with a solid chance of success.

Opportunities for ecological and economic improvement are immense, especially in the developing world. Since it is here that we find the most species of plants, the fewest botanists, and the greatest danger to the habitat, an important early step is to encourage study of the biological sciences and especially botany.

There are two key reasons why botany deserves such a high priority in a world where hunger threatens hundreds of millions of people. First, many potentially useful wild species may be lost as habitats are damaged. Second, yields of food crops can be increased through disease-resistant strains. For instance, a virus carried by the brown leafhopper recently destroyed vast areas of rice in Asia and the Pacific. Resistance to the plague was discovered in a few older varieties of rice from southern India and Sri Lanka and these types supplemented the plantings.

There are more than forty-seven thousand

At a site near Port-au-Prince, Haitians terrace and reforest denuded hillsides which held forests when Columbus discovered the island of Hispaniola in 1492. This project of the Inter-American Development Bank will reduce erosion and provide a measure of flood control.

varieties of cultivated rice—each carrying different genetic traits—in storage at the International Rice Research Institute (IRRI) in the Philippines. The varieties differ in such features as size, shape, disease and pest resistance, yield, grain size and other physical attributes, and in their efficient use of light in photosynthesis. The characteristics of each have been entered into computers so that particular types of rice can be matched to a grower's needs.

Other centers for preserving the genetic material of plants include The National Seed Storage Laboratory of the United States, established in 1958 at Fort Collins, Colorado. It now houses more than one hundred thousand samples of all sorts of seeds. There are wheat and corn banks in Mexico, a bean bank in Colombia, and banks for storing pearl millet and pigeon pea seeds in India. Approximately twelve thousand genetic stocks of seven potato species are maintained in the gene bank of the International Potato Center in Lima, Peru. Banana plants of many kinds have been collected in Southeast Asia for research and breeding. Wheat varieties have been assembled at the United States Department of Agriculture at Beltsville, Maryland. Another large seed collection is maintained at the Royal Botanic Gardens, Kew, where much of the practical methodology and physiological requirements for storing seeds were developed. These international seed banks are vital for the potential they hold for the world at large. They should always be supported by national seed banks to insure the plants' potential worth against power failure and to make them available for local breeding needs.

Botanical gardens of the world are an important facet of the conservation of genetic diversity. To assist in the documentation and propagation of new species, one hundred and

thirty have joined the Botanical Gardens Conservation Coordinating Body established under the auspices of the Threatened Plants Unit of the International Union for Conservation of Nature and Natural Resources (IUCN).

Many more botanical gardens need to be established in the developing world. Since tropical seeds are usually too large and moist to survive in present-day seed banks, live plantings in tropical settings are especially valuable. Support for existing botanical gardens is also essential since they provide the horticultural skills which will be needed for saving "wild" habitats and the species they contain.

Basic research into plant mechanisms can provide great economic leverage. One project of great potential resulted from the chance rediscovery of a "lost" species of teosinte, a form of wild corn, in 1978. Presumed extinct since 1910, *Zea diploperennis*, a perennial, was found growing on three acres of a hillside near Jalisco, Mexico. The plant population consisted of only one

or two thousand individuals, and they grew on land scheduled to be cleared to plant domesticated corn. Specimens were collected for testing and arrangements made to protect the wild plants. The teosinte was found to have twenty chromosomes, the same as corn (*Zea mays*), which makes hybridization theoretically possible. Because *Z. diploperennis* resists some diseases, one day its genes may contribute to varieties of corn which are disease resistant as well as perennial. Such an innovation would fit in with new efforts to introduce minimum-tillage agriculture to deter soil erosion. Had the Jalisco field been plowed up, humanity might have forever lost a remarkable opportunity.

The United States National Academy of Sciences (NAS) has brought together panels of experts to examine the genetic potential of many

Javanese children sell firewood from Calliandra calothyrsus, *a small South American tree with global potential. It quickly grows to a uniform size that can be felled with a machete.*

Windrows of brush protect Australian pine Casuarina esquisetifolia *seedlings planted along a beach in Niger. Established trees will bind the fragile soil, provide firewood and extract nitrogen for fertilizer from the air.*

wild and rare types of plants. Other countries around the world are also beginning to survey their own natural resources to find new or underexploited crop species. The first results are some fine publications which indicate how specific crops can open up new horizons within the developing world and thus demonstrate their importance to all mankind. The need and potential for worldwide conservation and economic development are indicated by a key passage from *Underexploited Tropical Plants with Promising Economic Value*, a NAS publication:

> *Most agricultural scientists are unaware of the scope and potential offered by tropical botany. The discipline suffers largely because the major centers of scientific research are located in temperate zones. There is an urgent need for plant researchers to become acquainted with tropical plant life. Important new products—such as oils, gums, and waxes for industry; proteins for food and feed; and chemicals for pest control—are likely to result from their attention.*

> *The variety of tropical plant species is staggering. Contained among them is a wealth of new products. In studying tropical economic botany it is not enough to consider solely traditional needs and markets. New raw materials also will be required in the future. Changing conditions are already creating demands for new*

products from previously under-exploited plants; more will be needed as pressures increase for the exploitation of renewable resources.

A companion volume, *Tropical Legumes*, addresses food species and also deals with the capacity of this botanical family to fix nitrogen from the air. To quote again:

> *... legumes are crucial to the balance of nature, for many are able to convert nitrogen gas from the air into ammonia, a soluble form of nitrogen, which is readily used by plants. While a few other plant families include species with this ability, legumes produce the great mass of biologically fixed nitrogen. . . . Even today, cultivated legume crops add more nitrogen to the soil worldwide than do commercial fertilizers.*

The NAS has also recommended that deforested areas in the humid tropics be replanted with several species of *Acacia, Calliandra, Derris, Gliricidia, Leucaena, Sesbania*, and other leguminous trees, all suppliers of their own fertilizer. The nitrogen-fixing properties of these species is especially vital in the many developing countries where commercial fertilizer is too expensive, and plantations of fast-growing trees will help replenish the firewood removed annually from jungle regions. Planting schemes employing native or suitable imported species could take pressure off virgin stands and save the soil. Since fertilizer is too expensive for general use in many developing countries, plants which produce their own are especially advantageous.

Research is still required before the correct tree species can be found for replanting in various areas. Some vigorously growing trees introduced under the wrong conditions may spread out of control to become weeds or otherwise create problems. Local species are usually better adapted to local ecological conditions, although they often seem to grow at intolerably slow rates. The Food and Agriculture Organization suggests breeding experiments, intensive studies of seed germination, and investigation of yield for prospective species. Only then can the genetic and biological attributes of trees suitable for semi-arid and arid areas become well enough known for confident planting in many regions.

One day it may be possible to transfer the genes permitting legumes to accommodate nitrogen-fixing bacteria on root nodules. Then other plants would be able to fix nitrogen, the most plentiful substance in the air but one which is difficult to extract. Since the nitrogen-fixing phenomenon is governed by a minimum of 17 genes, extensive research is likely to be required. In 1981, scientists transferred a gene from a French green bean into the cell of a sunflower seedling, having it "hitch a ride" on a bacterium injected into a chromosome of the sunflower. The resulting plant cells were called "sunbean" in honor of its parent species.

The International Rice Research Institute in the Philippines maintains a living library of rice varieties, below, to help protect the world's cultivated crop species. At right, an IRRI scientist examines a tiny rice flower spikelet.

BR 220-1-1
BANGLADESH

10th INTERNATIONAL RICE
YIELD NURSERY
(EARLY), 1982
ENTRIES: 27 (from 5 COUNTRIES)
D/S: SEPT. 29, 1982
D/P: OCT. 20, 1982
PLOT SIZE: 11.0 sq. m.
SPACING: 15 × 20 cm
LOCAL CHECK: IR50

CHIANUNG SEN YU 13
TAIWAN (CHINA)

Agroforestry

Much of the practical work in plant breeding is aimed at better agricultural production in tropical regions. A number of plant introductions have been created to support an especially promising technique, agroforestry. It stems from a form of home and village gardening practiced from antiquity. Studies and practical trials indicate that agroforestry and some other humble techniques are much more in harmony with the tropical environment than other of today's methods. In many areas, traditional agricultural expertise was forcibly displaced through the introduction of ambitious agricultural schemes—many of dubious worth outside of a colonial system—and is being threatened again as people of the less developed nations gravitate toward the cities.

Basically, in agroforestry, farmers and their families become partly foresters as well. They plant trees suitable for firewood production and intercrop them with their corn, bananas, fruit trees, spices, and beverage plants. Sometimes they also introduce domestic animals—anything from rabbits to water buffalo to crocodiles—for meat and fertilizer. Farm or village ponds can provide protein in the form of sunfish and other fast breeders.

While elements of such an approach have been in existence for centuries, the new initiatives aim at extending and refining the process. Small plots of food or cash crops and woodlots can be integrated into existing fields and paddy layouts in ways that increase overall yields while reducing damage to the soil.

Agroforestry and its variations do not represent a revolutionary departure from familiar agricultural techniques—only a change in emphasis and partial return to a conservative approach involving the labor of every family member. Such an approach, administered at the local level, can have great survival value.

One practical example of agroforestry involves a tree mentioned in the earlier chapter on underexploited plants. The foliage of the leguminous *Acacia albida* grows during the season when most other trees lose their leaves, allowing crops planted between the trees to receive full sun during their growing season. While the soil does double duty, it is fertilized twice: once by the nitrogen fixed in its roots and again with the manure from animals that eat the tree leaves in the dry season. Local observers believe that the tree develops a symbiotic relation with the other

Mohandas K. Gandhi sought guidance from George Washington Carver on a vegetarian diet that would provide all essential nutrients. Both men looked to "poor man's foods" to assure the well-being of rural populations in their homelands. To this end, Carver (shown on the stamp below) taught his fellow Southerners how to restore the vitality of exhausted soil and promoted widespread cultivation of peanuts, sweet potatoes, and black-eyed peas. Gandhi's statue bears a loop of marigolds, India's national flower.

231

plants which grow near it. A Sultan of Zinder, in what is now Niger, deemed this acacia sacred and cut off the hands of anyone who destroyed a tree.

From Bali to the Sahel to Amazonia, the benefits of agroforestry are manifold. In South America's jungle, intercropping of black pepper with corn, fruit, vegetables, and fodder plants is gaining acceptance with agroforesters. The soil is protected even as a cash crop grows. Asians have long realized the value of intercropping, with some of the most successful applications

found on the Indonesian island of Bali. The Sahel once held gardens noted for their figs, grapes, olives, and even oranges, lemons, and apricots. It is clear that an abundance of plants on the land reduces the run-off of rain and helps to hold moisture in the soil. In fact, reforestation has even caused dried-up springs to start running again. Moreover, some climatologists suggest that greater amounts of green on the ground, especially forest cover, may even increase levels of rainfall year in and year out. It has been suggested that trees, perhaps through

Bali is an Indonesian island with an ancient heritage from India. Intensive land use permits a finite area to support a growing population. Local tradition has incorporated forestry and farming into a unified system greatly admired by the authors. Left, papaya grows between the rice fields. Below, an ornate scarecrow wards off wild birds from a paddy field even as domesticated ducks strip insects from the stubble. The ducks provide eggs and meat, and the cleaned field is less likely to be infested with pests and disease.

their transpiration of water from the ground, somehow induce or attract rainfall.

Scope of Our Action

In much of the developing world, conservation is intimately involved with the survival of the people themselves. If success is to be widespread, commitment to measures such as reforestation and the introduction of agroforestry at the local level must be supported by programs at the national level as well as by cooperation among developing and developed nations.

Conservation for development thus takes on new meaning, and recognition of the need for a comprehensive global approach to conservation is a natural outcome.

Perhaps we are finally witnessing something new under the sun. The World Conservation Strategy, published in March of 1980, was produced by the International Union for Conservation of Nature and Natural Resources (IUCN) with the cooperation and financial support of the United Nations Environmental Programme (UNEP) and the World Wildlife Fund (WWF).

The Strategy was prepared in collaboration with the Food and Agricultural Organization (FAO) and the United Nations Educational, Scientific, and Cultural Organization (Unesco).

As we see it, The World Conservation Strategy illuminates the means by which conservation, development, and the aims and hopes of people in all parts of this planet can be brought into a harmonious plan for the future. It is a tribute to its authors and collaborators—the voluntary conservation movement in many countries, scientific researchers, governmental and international agencies—that such an essential and widely representative document could come into being.

As indicated in the Strategy, close consultation between outside scientists and local people can achieve novel and profitable schemes for economic development and biological conservation. Some of these can enhance agroforestry. For instance, tribesmen of Papua New Guinea had long known the market value of rare and beautiful butterflies in their area, with some prime specimens worth a thousand dollars in 1980. Today, the local people have learned how to build butterfly farms, enclosures planted with the insects' favorite food plants. The farmers protect the eggs, larvae, chrysalises and the emerging butterflies. Some are taken for specimens and sold while the breeding stock remains safe. This supplements crop production and provides a valuable additional cash income.

The late René Dubos, a noted biologist and conservationist, suggested that man's influence on the land can be either a curse or a blessing. Dubos pointed to the productive heartland of his native France as an example of ground which was improved and even glorified through human toil. Here, as in almost every locality in the world and with nearly all efforts to combine production with conservation, local conditions often can be carefully exploited by the local inhabitants for their own sustainable profit and for the benefit of all those with whom they trade and deal. In this lesson perhaps lies the greatest wisdom we can use to save our green world.

Flinging water is a routine farm chore in Thailand that becomes a festive observance at New Years. The government is seeking agricultural reform on the local level, in part to discourage opium traffic in the north.

PART V

A Third Chance

MONUMENTS TO PAST GLORY RISE IN EVERY land, as with the Taj Mahal in northern India. Storytellers of ancient Sumer though, perhaps best caught mankind's universal sense of lost joys and opportunities in their tales of Eden. On history's pages, the first major loss of garden space, and the privilege associated with it, probably happened in the classical Mediterranean, Mesopotamia, and parts of the Middle East. That so much beauty and vitality can survive amid such deeply scarred landscapes is remarkable and encouraging.

Columbus discovered one solution to the problem of rising poverty at home, an entire "new world" nearly as big as the old one. During only five hundred years, though, some of the Americas have come to support greater poverty and environmental damage than the Old World. Even in the last century, a good many Europeans by-passed the Americas for a chance at heaven-on-earth in the South Pacific.

Though there is nowhere else to go, geographically speaking, there is hope for a real "third chance." The Green Revolution of the 1960s and seventies bought time, and now the economic botanists are regrouping for an even more efficient assault on hunger and disease. In the following Epilogue, Mrs. Indira Gandhi emphasizes the importance of the truly international character of any new undertaking. To succeed, such shared effort would require levels of cooperation and shared vision that have rarely been attained during the first eight decades of the twentieth century.

Finally, the authors gratefully acknowledge the source of the title of this volume. During a long search, several titles were tried. Finally, in Mrs. Gandhi's epilogue, the perfect words appeared. Though now used elsewhere in the book, the phrase "green and living world" originated with Mrs. Indira Gandhi.

Epilogue— Goals for the New Century

Mrs. Indira Gandhi

In India, nimble goats contrast with stalwart zebu cattle whose mural procession they parallel, above. In an Indian market opposite, a young boy surveys an array of produce, including both tropical mangoes, and gourds, and temperate-climate vegetables such as carrots and tomatoes. Also for sale are potatoes originally from the Andes of South America that grow well in India's highlands.

FOR COUNTLESS CENTURIES INDIAN CIVILIZA-tion has proclaimed the oneness of all existence and the unity of life and non-life. This insight is borne out by modern scientific discovery. The destruction of any one form of life inexorably affects other forms of life, for each has a purpose and contributes to the balance of the whole. Ultimately the survival of the human race will depend on how well we can maintain this balance.

It is ironical that at the very stage when this ancient tenet of Indian philosophy is being recognized by the world, our own people are allowing deforestation and the extinction of various species. The compulsions of development, the pressures of population, and the greed for profit have all combined to threaten our forests, our animals and the very air that we breathe.

Big projects and modern developments do not hold all the answers. Through our long history, village communities have often evolved productive partnerships with nature. Scientists are now recognizing the sound common sense of their ways—for instance the planting of many crop species within limited areas. This provides a barrier to the spread of plant pests and diseases. The resulting variety of foodstuffs also induces public health through better nutrition while lowering costs through local supply. In another example, many of our cattle breeds are robust and quite resistant to disease. We must thank the villagers who cross domestic animals with semi-wild types either caught in the forest or kept for that purpose.

There is not and has never been any contradiction between conservation and development. The two must go together. Development is distorted without conservation. Modern science has drawn attention to the genetic values of wild plants and animals. In the true sense productivity is possible only in a clean environment with a healthy population. But the temptations of short-term gains overshadow a long-term view. Such negligence can be disastrous. How much longer can our earth withstand the exploitation to which it has been increasingly subjected since the advent of the industrial revolution? We must restore the strength that has been sapped from the environment. The green world can regain the ability to renew itself. We must make it capable of supporting an ever-growing population with consequently increasing demands on natural resources.

It is our country's Natural Forest Policy to afforest 33 percent of our total land area. Ac-

tually only 26.8 percent has woodland cover. So we have launched a number of schemes to try and prevent the forest cover from being further depleted. The wealth of forests cannot be measured as revenue but in terms of the nation's well-being. We have banned the felling of trees beyond an altitude of 1,000 metres. Our Forest Conservation Act disallows the use of forest for any purpose whatever without the permission of the central government. Since 1980 we have begun a vigorous afforestation programme, which includes social and farm forestry and development of alternative sources of energy. Wood is being used for cooking, heating in the cold of our northern winter, and for paper and other industries. Social forestry programmes provide timber for fuel and our work for alternative sources of energy is directed specially to reduce the pressure on wood.

Already the world and my own country have lost a great deal by neglecting the ecological equilibrium. Little time is now left to us. Yet it is not too late to make amends. The world-wide awareness of the environment is a welcome development. Modern technology enables us to initiate programmes which may not have been possible earlier. The successful work in the United Kingdom of taking lands damaged by strip mining and turning them into golf courses, parklands and forests is an example to the rest of the world. It also represents the best use of creative technology.

Human beings can thrive only in a green and living world. The problems of the environment have to be faced by the world as one. Fragmentation in our thinking and doing, or piecemeal treatment, will not serve our purpose. India is in support of any movement for conservation on a global basis.

After the harvest, in which rice tassels are clipped by hand, these women in India gather the straw for thatch, fodder, and fuel.

Acknowledgments

The Authors

In addition to their scholarly accomplishments, the four authors are practical men with first-hand knowledge of the world gained through extensive travel. Seldom equalled, their experience allows for a remarkably authoritative view of our living planet.

Professor Edward S. Ayensu is Secretary General of the International Union of Biological Sciences. A citizen of Ghana, and one of Africa's leading scientists, he is particularly active in international conservation through his work at the Smithsonian Institution, where he is a Senior Botanist in the Department of Botany. A prolific author, Professor Ayensu has also made many television appearances and is active in fund-raising for scientific research. He is recognized as an international authority on medicinal plants and traditional systems of health-care.

Professor Vernon H. Heywood, a scientist with an immense international scientific reputation, is head of the Department of Botany at the University of Reading, England. In recent years, he has devoted much of his attention to the problems of plant conservation, particularly in the developing regions of the New World. His contributions to the literature include numerous scientific books and articles as well as volumes for a general audience on plants and their place in global ecology.

Grenville L. Lucas, O.B.E., is Acting Keeper of the Herbarium of the Royal Botanic Gardens, Kew and has received the O.B.E. for service to conservation. He is a trustee of the World Wildlife Fund (U.K.), Chairman of the IUCN's Species Survival Commission, and co-author of *The IUCN Plant Red Data Book.* Mr. Lucas is probably the United Kingdom's foremost plant conservationist.

Robert A. DeFilipps, a United States citizen, is a museum specialist with the Plant Conservation Unit, Department of Botany, United States National Museum of Natural History. He has worked on aspects of world rare and endangered plant species for a decade, and presently devotes much time to the development of biogeographical archives for the documentation of ecological changes on the world's tropical islands. He has participated in the *Flora Europaea* and *Flora of Dominica* projects, and is a member of The West India Committee (London), and the International Palm Society.

Distinguished Contributors

S. Dillon Ripley, a noted ornithologist, became the eighth Secretary of the Smithsonian Institution in February 1964.

HRH The Prince Philip, for many years a respected conservationist, became the third president of the International World Wildlife Fund in 1981.

Mrs. Indira Gandhi is Prime Minister of India and a leading advocate of international cooperation for sustainable development.

With Special Thanks

Arthur E. Bell—Director, Royal Botanic Gardens. Kew

T.N. Khoshoo—Secretary for the Environment, Government of India, New Delhi

John E. McCarthy—President, International Catholic Migration Commission, Geneva

Kenton R. Miller—Director General, International Union for Conservation of Nature and Natural Resources

Noel Vietmeyer—National Academy of Sciences, Washington, D.C.

Deepak Vohra—First Secretary of Press, Indian Embassy, Washington, D.C.

Pat and Ted Vosburgh—Bethesda, Maryland

Sun Weixue—Second Secretary (Cultural) of the Embassy of the People's Republic of China in the United States of America

---

Kate Alfriend—Office of Information, USDA

Gus Van Beek—Curator, Department of Anthropology, National Museum of Natural History

Iris Bradshaw—Inter-American Development Bank

Valerie Brown—Free-lance Photography, Washington, D.C.

James R. Buckler—Director, Office of Horticulture, Smithsonian Institution

Ed Castle—Free-lance Photography, Chevy Chase, Maryland

Smithsonian Institution photographer Kjell B. Sandved seeks details of plant pollination by insects in a Maryland wetland.

Janet Cavanaugh—George Peabody Library of The John Hopkins University

Anne F. Clapp—Paper Conservator

Anni-Siranne Coplan—Free-lance Photography, Washington, D.C.

Spencer R. Crew—Historian, Archives Center, National Museum of American History

José Cuatrecasas—Research Associate, Department of Botany, National Museum of Natural History

Tom DeClaire—Geography and Map Division, Library of Congress

Mark Dimmitt—Botanist, Arizona-Sonora Desert Museum, Tucson, Arizona

Paul Fonseca, World Bank

F. Raymond Fosberg—Botanist Emeritus, Department of Botany, National Museum of Natural History

Bryna Freyer—Archives, National Museum of African Art

Fritz Frommeyer—Media Relations, Eli Lilly and Company

Patricia Geeson—Museum Technician, National Museum of American Art

Toni Warner Gillas—Free-lance Indexer

Howard S. Golden—Publications Officer, Publications Services Branch, NASA Headquarters

Joe A. Goulait—Color Lab Chief, Museum of Natural History Branch, Smithsonian Office of Printing and Photographic Services

Christopher Grey-Wilson—Royal Botanic Gardens, Kew

Charles R. Gunn—Plant Exploration and Taxonomy Laboratory, Agricultural Research Service, USDA

John Hoke—National Capitol Region, United States Department of the Interior

Francis M. Hueber—Curator, Department of Paleobiology, National Museum of Natural History

Carole Jacobs—Free-lance Writer Editor

Alison Jolly—The Rockefeller University, New York City

Nancy Knight—Museum Specialist, Division of Medical Sciences, National Museum of American History

Victor E. Krantz—Chief, Museum of Natural History Branch, Smithsonian Office of Printing and Photographic Services

Jane Lamlein—Plant Conservation Unit, Department of Botany, National Museum of Natural History

David B. Lellinger—Associate Curator, Department of Botany, National Museum of Natural History

Mark M. Littler—Chairman, Department of Botany, National Museum of Natural History

Mary Mangone—The Smithsonian Office of Educational Research

David Mangurian—Inter-American Development Bank

Brian Mathew—Royal Botanic Gardens, Kew

Tim McCabe—Photographer, Soil Conservation Service, USDA

Charles R. McClain—Oceans and Ice Branch, Laboratory for Atmospheric Sciences, Goddard Space Flight Center, NASA

David McGlyment, Department of Egyptian Antiquities, British Museum

Jean Miller—Folger Shakespeare Library

Max E. Miller—Earth Satellite Corporation

Jane Montant—*Gourmet* Magazine

Robert Newton—World Bank

Joan W. Nowicke—Curator, Department of Botany, National Museum of Natural History

Jerry Olson—Oak Ridge National Laboratories

Alden Prouty—Manager, Marketing Communications Services, St. Regis Corporation

Enayetur Rahim—Library Technician, Smithsonian Institution Libraries

Ruth Reck—Physics Department, General Motors Research Laboratories

Jonathan Reel—Archives, National Museum of African Art

Philip Ross—National Academy of Sciences, Washington, D.C.

Anna M. Rowsell—Cultural Documentation, United Kingdom

Frances Rowsell—Graphics Research, United Kingdom

Marie-Hélène Sachet—Associate Curator, Department of Botany, National Museum of Natural History

Ruth F. Schallert—Assistant Natural History Librarian, National Museum of Natural History

Alfred H.M. Shehab—Office of Public Affairs, Goddard Space Flight Center, NASA

Nancy Tosta-Miller—Land and Water Use Analyst, Department of Forestry, State of California

Charles Vermillion—Program Manager for Direct Readout Application, Goddard Space Flight Center, NASA

James H. Wallace, Jr.—Director, Museum of Natural History Branch, Smithsonian Office of Printing and Photographic Services

Ellen B. Wells—Chief, Special Collections, Smithsonian Institution Libraries

James White—Hunt Institute For Botanical Documentation, Pittsburgh

John P. Wiley, Jr.—Member, Board of Editors, *Smithsonian* Magazine

---

June G. Armstrong—Production Assistant, Direct Mail Book Division, Smithsonian Institution Press

Kathleen V.W. Brown—Production Assistant, Production Section, SIP

Deborah E. Corsi—Assistant Editor, SIP

Maureen R. Jacoby—Managing Editor, SIP

Lawrence J. Long—Production Manager, SIP

John F. Ross—Free-lance Editor and Researcher

Patricia Upchurch—Picture Editor, Direct Mail Book Division, SIP

World Conservation Strategy

The World Conservation Strategy was published through the combined efforts of the International Union for Conservation of Nature and Natural Resources (IUCN), the United Nations Environment Programme (UNEP), and the World Wildlife Fund (WWF) in 1980.

Two critical themes used throughout the Strategy—conservation and development—are defined as follows:

Conservation—the management of human use of the biosphere so that it may yield the greatest sustainable benefit to present generations while maintaining its potential to meet the needs and aspirations of future generations.

Development—the modification of the biosphere and the application of human, financial, living and non-living resources to satisfy human needs and improve the quality of human life.

The Executive Summary of the World Conservation Strategy is reproduced below in facsimile.

Executive Summary

The World Conservation Strategy is intended to stimulate a more focussed approach to the management of living resources and to provide policy guidance on how this can be carried out by three main groups:
— *government policy makers and their advisers;*
— *conservationists and others directly concerned with living resources;*
— *development practitioners, including development agencies, industry and commerce, and trade unions.*

1. The aim of the World Conservation Strategy is to achieve the three main objectives of living resource conservation:
a. **to maintain essential ecological processes and life-support systems** (such as soil regeneration and protection, the recycling of nutrients, and the cleansing of waters), on which human survival and development depend;
b. **to preserve genetic diversity** (the range of genetic material found in the world's organisms), on which depend the functioning of many of the above processes and life-support systems, the breeding programmes necessary for the protection and improvement of cultivated plants, domesticated animals and microorganisms, as well as much scientific and medical advance, technical innovation, and the security of the many industries that use living resources;
c. **to ensure the sustainable utilization of species and ecosystems** (notably fish and other wildlife, forests and grazing lands), which support millions of rural communities as well as major industries.

2. These objectives must be achieved as a matter of urgency because:
a. **the planet's capacity to support people is being irreversibly reduced in both developing and developed countries:**
— thousands of millions of tonnes of soil are lost every year as a result of deforestation and poor land management;
— at least 3,000 km² of prime farmland disappear every year under buildings and roads in developed countries alone;
b. **hundreds of millions of rural people in developing countries, including 500 million malnourished and 800 million destitute, are compelled to destroy the resources necessary to free them from starvation and poverty:**
— in widening swaths around their villages the rural poor strip the land of trees and shrubs for fuel so that now many communities do not have enough wood to cook food or keep warm;
— the rural poor are also obliged to burn every year 400 million tonnes of dung and crop residues badly needed to regenerate soils;
c. **the energy, financial and other costs of providing goods and services are growing:**
— throughout the world, but especially in developing countries, siltation cuts the lifetimes of reservoirs supplying water and hydroelectricity, often by as much as half;
— floods devastate settlements and crops (in India the annual cost of floods ranges from $140 million to $750 million);
d. **the resource base of major industries is shrinking:**
— tropical forests are contracting so rapidly that by the end of this century the remaining area of unlogged productive forest will have been halved;

— the coastal support systems of many fisheries are being destroyed or polluted (in the USA the annual cost of the resulting losses is estimated at $86 million).

3. *The main obstacles to achieving conservation are:*
a. **the belief that living resource conservation is a limited sector**, rather than a process that cuts across and must be considered by all sectors;
b. **the consequent failure to integrate conservation with development;**
c. **a development process that is often inflexible and needlessly destructive**, due to inadequacies in environmental planning, a lack of rational use allocation and undue emphasis on narrow short term interests rather than broader longer term ones;
d. **the lack of a capacity to conserve**, due to inadequate legislation and lack of enforcement; poor organization (notably government agencies with insufficient mandates and a lack of coordination); lack of trained personnel; and a lack of basic information on priorities, on the productive and regenerative capacities of living resources, and on the trade-offs between one management option and another;
e. **the lack of support for conservation**, due to a lack of awareness (other than at the most superficial level) of the benefits of conservation and of the responsibility to conserve among those who use or have an impact on living resources, including in many cases governments;
f. **the failure to deliver conservation-based development where it is most needed**, notably the rural areas of developing countries.

4. *The World Conservation Strategy therefore:*
a. **defines living resource conservation and explains its objectives**, its contribution to human survival and development and the main impediments to its achievement (sections 1-4);
b. **determines the priority requirements for achieving each of the objectives** (sections 5-7);
c. **proposes national and subnational strategies** to meet the priority requirements, describing a framework and principles for those strategies (section 8);
d. **recommends anticipatory environmental policies, a cross-sectoral conservation policy and a broader system of national accounting** in order to integrate conservation with development at the policy making level (section 9);

e. **proposes an integrated method of evaluating land and water resources, supplemented by environmental assessments,** as a means of improving environmental planning; and **outlines a procedure for the rational allocation of land and water uses** (section 10);
f. **recommends reviews of legislation** concerning living resources; **suggests general principles for organization within government;** and in particular **proposes ways of improving the organizational capacities for soil conservation and for the conservation of marine living resources** (section 11);
g. **suggests ways of increasing the number of trained personnel;** and **proposes more management-oriented research and research-oriented management**, so that the most urgently needed basic information is generated more quickly (section 12);
h. **recommends greater public participation** in planning and decision making concerning living resource use; and **proposes environmental education programmes and campaigns** to build support for conservation (section 13);
i. **suggests ways of helping rural communities to conserve** their living resources, as the essential basis of the development they need (section 14).

5. *In addition, the Strategy recommends international action to promote, support and (where necessary) coordinate national action, emphasizing in particular the need for:*
a. **stronger more comprehensive international conservation law,** and **increased development assistance for living resource conservation** (section 15);
b. **international programmes** to promote the action necessary to conserve **tropical forests and drylands** (section 16), to protect areas essential for the preservation of **genetic resources** (section 17), and to conserve the global "commons"—**the open ocean, the atmosphere, and Antarctica** (section 18);
c. **regional strategies** to advance the conservation of **shared living resources** particularly with respect to **international river basins and seas** (section 19).

6. *The World Conservation Strategy ends by summarizing* **the main requirements for sustainable development,** indicating conservation priorities for the Third Development Decade (section 20).

The WWF and IUCN (1984-85)

WWF (World Wildlife Fund) is a private, international conservation foundation based in Switzerland. Founded in 1961 for the conservation of nature and the natural environment, the WWF has channelled over $85 million into more than 3,800 projects in some 130 countries.

IUCN (International Union for Conservation of Nature and Natural Resources) is a network of governments, nongovernmental organizations (NGOs), scientists, and other conservation experts, joined together to promote the protection and sustainable use of living resources. Founded in 1948, the IUCN has more than 450 member governments and NGOs in over 100 countries.

The WWF and IUCN maintain a joint headquarters in Switzerland at the following address:

World Conservation Centre
Avenue du Mont-Blanc
1196 Gland, Switzerland
telephone (022) 647181
telex: 28183 (WWF); 22618 (IUCN)

Following are listed the addresses of the WWF's twenty-four worldwide affiliates, two WWF organizational links, and three IUCN Outstations.

Editor's Note:

The Executive Summary of the World Conservation Strategy appears on pages 244-245 by permission of IUCN. The entire document is most highly recommended, and represents a remarkable synthesis and consensus by hundreds of contributors from many lands. It may be obtained through the addresses listed here, and through some United Nations sources.

WWF Affiliate Organizations

WWF-AUSTRALIA
Level 17 St. Martins Tower
31 Market Street
Sydney
NSW 2000
Tel: (02) 29 1602

WWF-AUSTRIA
(Oesterreichischer Stiftverband für Naturschutz)
Ottakringer Str. 120
Postfach 1
A-1162 Vienna
Tel: (0222) 46 14 63

WWF-BELGIUM
937 Chaussée de Waterloo B5
B-1180 Brussels
Tel: (02) 375 3498

WWF-CANADA
60 St. Clair Av. East
Suite 201
Toronto, Ontario M4T IN5
Tel: (416) 923 8173

WWF-DENMARK
(Verdensnaturfonden)
H.C. Andersens Boulevard 31
DK-1553 Copenhagen V
Tel: (01) 13 20 33

WWF-FINLAND
(Maailman Luonnon Säätiö Suomen Rahasto)
Uudenmaankatu 40
SF-00120 Helsinki 12
Tel: 644.511

WWF-FRANCE
(Association Française du World Wildlife Fund)
14 rue de la Cure
F-75016 Paris
Tel: (01) 527 86 76

WWF-GERMANY
(WWF-Deutschland)
Bockenheimer Anlage 38
D-6000 Frankfurt am Main
Tel: (0611) 72 51 55

WWF-HONG KONG
4th Floor
Wing On Life Building
22, Des Voeux Rd. C
Hong Kong
Tel: (05) 264473

WWF-INDIA
c/o Godrej & Boyce Mfg. Co.
Private Ltd.
Lalbaug, Parel
Bombay 400012
Tel: 44 13 61

WWF-ITALY
(Associazione Italiana per il
World Wildlife Fund)
Via P.A. Micheli 50
I-Rome 00197
Tel: (06) 80 20 08

WWF-JAPAN
5F Yamaki Building
Sotokanda 4-8-2
Chiyoda-ku
Tokyo 101
Tel: (03) 255 3770

WWF-LUXEMBOURG
Musée d'Histoire Naturelle
Marché aux Poissons
L-Luxembourg
Tel: 47 87 20

WWF-MALAYSIA
P.O. Box 769
Kuala Lumpur
Tel: (03) 945777

WWF-NETHERLANDS
(Wereld Natuur Fonds)
Postbus 7
NL-3700 AA Zeist
Tel: (03404) 22.164

WWF-NEW ZEALAND
P.O. Box 12-200
Wellington North
Tel: 851-389

WWF-NORWAY
(Verdens Villmarksfond)
Övre Slottsgate 7
N-Oslo 1
Tel: (02) 42 43 15

WWF-PAKISTAN
P.O. Box 1312
Lahore
Tel: 853.062

WWF-SOUTH AFRICA
(S.A. Nature Foundation)
P.O. Box 456
7600 Stellenbosch
Tel: (02231) 72892/3

WWF-SPAIN
(Asociacion para la Defensa
de la Naturaleza - ADENA)
6 Santa Engracia
E-Madrid 10
Tel: (01) 410 2401

WWF-SWEDEN
(Världsnaturfonden)
Fituna
S-140 41 Sorunda
Tel: (0753) 44 143

WWF-SWITZERLAND
Postfach CH-8037
Zürich
Tel: (01) 44 20 44

WWF-UNITED KINGDOM
Panda House
11-13 Ockford Road
Godalming, Surrey GU7 1QU
Tel: (04868) 20.551

WWF-UNITED STATES
(World Wildlife Fund Inc.)
1601 Connecticut Ave. N.W.
Washington, D.C. 20009
Tel: (202) 387-0800

Other Organizational Links
with WWF

WWF-CHINA JOINT COMM.
c/o The Environmental
Protection Office of the
State Council
Beijing, China

VENEZUELA
(Fundacion para la Defensa
de la Naturaleza - FUDENA)
Apartado 70376
Caracas 107
Tel: 239.41.11
      239.45.10

IUCN Outstations

*(For the directorate, and sections
on fauna and wildlife trade)*
IUCN Conservation Monitoring
Centre
219c Huntingdon Road
Cambridge CB3 ODL, U.K.
Tel: (0223) 277314/277420

*(For the computer support group
and sections on plants and
protected areas)*
The Herbarium
Royal Botanic Gardens
Kew, Richmond
Surrey TW9 3AB, U.K.
Tel: (01) 940-1171

IUCN Environmental Law
Centre

Adenauerallee 214
D-5300 Bonn
Federal Republic of Germany
Tel: (02 28) 21 34 52

Suggested Readings

Ayensu, E.S., ed. *Jungles.* New York: Crown Publishers, 1980.

————. and R.A. DeFilipps. *Endangered and Threatened Plants of the United States.* Washington, D.C.: Smithsonian Institution and World Wildlife Fund-U.S., 1978.

Bannerman, R.H., Burton, J. and C. Wen-Chieh. *Traditional Medicine and Health Care Coverage.* Geneva: World Health Organization, 1983.

Blunt, W. *The Compleat Naturalist—A Life of Linnaeus.* London: William Collins, Sons & Company Ltd., 1971.

Botting, D. *Humboldt and the Cosmos.* London: Michael Joseph Ltd., 1973.

Bramwell, D., ed. *Plants and Islands.* London: Academic Press, 1979.

Carlquist, S. *Island Life.* New York: Natural History Press, 1965.

Carr, A. *The Everglades—The American Wilderness.* New York: Time-Life Books, 1973.

Chapman, V. J., ed. *Wet Coastal Ecosystems—Ecosystems of the World 1.* New York: Elsevier Scientific Publishing Co., 1977.

Cloudsley-Thompson, J. and E. Duffey. *Deserts and Grasslands: The World's Open Spaces.* Garden City, New York: Doubleday and Company, Inc., 1977.

Frankel, O.H. and J. G. Hawkes, eds. *Crop Genetic Resources for Today and Tomorrow.* New York: Cambridge University Press, 1975.

Giddy, C. *Cycads of South Africa.* Cape Town: Purnell, 1974.

Goodin, J.R. and D.K. Northington, eds. *Arid Land Plant Resources.* Lubbock, Texas: Texas Tech University, 1979.

Harrison, S.G., Masefield, G.B., Nicholson, B.E. and M. Wallis. *The Oxford Book of Food Plants.* London: Oxford University Press, 1969.

Hepper, F.N., ed. *Kew—Gardens For Science and Pleasure.* London: Her Majesty's Stationery Office, 1982.

Heywood, V.H., ed. *Flowering Plants of the World.* New York: Mayflower Books, 1978.

Hoke, J. *Ecology.* New York: Franklin Watts, 1977.

Horwitz, E.L. *Our Nation's Wetlands.* Washington, D.C.: Council on Environmental Quality, 1978.

Hulton, P. and L. Smith. *Flowers in Art from East and West.* London: British Museum Publications, Ltd., 1979.

Huxley, A. *Green Inheritance—The World Wildlife Fund Book of Plants.* London: Collins/Harvill, 1984.

Jolly, A. *A World Like Our Own: Man and Nature in Madagascar.* New Haven: Yale University Press, 1980.

Ketchum, R. M. *The Secret Life of The Forest.* New York: American Heritage Press, 1970.

Koopowitz, H. and H. Kaye. *Plant Extinction: A Global Crisis.* Washington, D.C.: Stone Wall Press, 1983.

Little, E.L., Jr. *Common Fuelwood Crops: A Handbook for Their Identification.* Morgantown, W.V.: Communi-Tech Associates, 1983.

Lucas, G. and H. Synge. *The IUCN Plant Red Data Book.* Morges, Switzerland: IUCN, 1978.

Marsh, G.P. *Man and Nature.* Edited by D. Lowenthal. Cambridge: The Belknap Press of Harvard University Press, 1965.

Mohlenbrock, R.H. *Where Have All The Wildflowers Gone?* New York: Macmillan Publishing Co., Inc., 1983.

Moore, D.M., ed. *Green Planet: The Story of Plant Life on Earth.* Cambridge: Cambridge University Press, 1982.

Moorehead, A. *Darwin and The Beagle.* London: Book Club Associates, 1969.

Mountfort, G. *Portrait of a Wilderness: The Story of the Coto Doñana Expeditions.* London: Hutchinson, 1958.

Myers, N. *Conversion of Tropical Moist Forests.* Washington, D.C.: National Academy of Sciences, 1980.

————. *The Primary Source: Tropical Forests and Our Future.* New York: Norton, 1984.

National Academy of Sciences. *Underexploited Tropical Plants With Promising Economic Value.* Washington, D.C.: National Academy of Sciences, 1975.

————. *The Winged Bean: A High-Protein Crop for the Tropics.* Washington, D.C.: National Academy of Sciences, 1975.

————. *Leucaena: Promising Forest and Tree Crop for the Tropics.* Washington, D.C.: National Academy of Sciences, 1977.

————. *Firewood Crops: Shrub and Tree Species for Energy Production.* Washington, D.C.: National Academy of Sciences, 1980.

National Research Council. *Products From Jojoba: A Promising New Crop For Arid Lands.* Washington, D.C.: National Academy of Sciences, 1975.

————. *Tropical Legumes: Resources for the Future.* Washington, D.C.: National Academy of Sciences, 1979.

————. *Agroforestry in the West African Sahel.* Washington, D.C.: National Academy Press, 1983.

————. *Genetic Engineering of Plants.* Washington, D.C.: National Academy Press, 1984.

Ponnamperuma, C. *The Origins of Life.* London: Thames and Hudson, 1972.

Prance, G.T. and T.S. Elias, eds. *Extinction is Forever.* Bronx, N.Y.: New York Botanical Garden, 1977.

Rowley, G. *The Illustrated Encyclopedia of Succulents.* New York: Crown Publishers, 1978. (IOS Code of Conduct, see p. 211)

Schery, R.W. *Plants For Man.* Englewood Cliffs, N.J.: Prentice-Hall, 1972.

Schultes, R.E. and A. Hofmann. *Plants of the Gods: Origins of Hallucinogenic Use.* New York, St. Louis and San Francisco: McGraw-Hill Book Company, 1979.

Sheffield, C. *Earth Watch: A Survey of the World From Space.* New York: Macmillan, 1981.

Simmons, J.B., Beyer, R.I., Brandham, P.E., Lucas, G.L., and V.T.H. Parry, eds. *Conservation of Threatened Plants.* New York: Plenum Press, 1976.

Synge, H., ed. *The Biological Aspects of Rare Plant Conservation.* New York: John Wiley & Sons, 1981.

Thesiger, W. *The Marsh Arabs.* New York: Dutton, 1964.

U.S. Department of Agriculture. *1983 Yearbook of Agriculture, Using Our Natural Resources.* Washington, D.C.: U.S. Government Printing Office, 1983.

Ward, N.B. *On The Growth of Plants in Closely Glazed Cases.* London: John Van Voorst, Paternoster Row, 1852.

Waksman, S.A. *Humus: Origin, Chemical Composition, and Importance in Nature.* Baltimore: The Williams & Wilkins Company, 1936.

Whitehead, P. *The British Museum (Natural History).* London: Scala/Philip Wilson in association with the British Museum (Natural History), 1981.

Westoby, J.C. *World Forest Development: Markets, Men and Methods.* Vancouver: University of British Columbia, 1965.

Wilson, C.M. *Trees and Test Tubes: The Story of Rubber.* New York: Henry Holt and Company, 1943.

Periodicals

Ames, B.N. "Dietary Carcinogens and Anticarcinogens." *Science* 221 (4617) (1983): 1256-1264.

Amos, W.H. "The Estuary." *Defenders, Educational Supplement* 55-4-a (1980): 225-240.

Ashton, P.S. "Speciation in the Tropical Rain Forest: Where Do We Stand Now?" *Australian Society for Systematic Botany* 28 (1981): 27-31.

Ayensu, E.S. "A Worldwide Role for the Healing Powers of Plants." *Smithsonian* 12 (8) (1981): 86-97.

Betz, R.F. "Resurrecting the Prairie." *Garden* 3 (4) (1979): 28-34.

Campbell, F.T. "Controlling the Trade in Plants: A Progress Report." *Garden* 7 (4) (1983): 2-5, 32.

Canby, T.Y. "El Nino's Ill Wind." *National Geographic* 165 (2) (1984): 144-183.

Christensen, B. "Mangroves—What Are They Worth?" *Unasylva* 35 (139) (1983): 2-15.

Davenport, W. "The Camargue: France's Wild, Watery South." *National Geographic* 143 (5) (1973): 696-726.

Davis, E. W. "Preparation of the Haitian Zombie Poison." *Harvard University, Botanical Museum Leaflets* 29 (2) (1983): 139-149.

DiCastri, F. and G. Glaser. "Highlands and Islands: Ecosystems in Danger." *Unesco Courier* (April 1980): 6-11.

Egge, R.E. "Martius, the Father of Palms." *Principes* 23 (4) (1979): 158-170.

Farnsworth, N.R. "How Can The Well Be Dry When It Is Filled With Water?" *Economic Botany* 38 (1) (1984): 4-13.

Fosberg, F.R. and M.-H. Sachet. "Henderson Island Threatened." *Environmental Conservation* 10 (2) (1983): 171-173.

Fulder, S. "The Drug That Builds Russians." *New Scientist* 87 (1980): 576-579.

Garden. "Special Issue: The Amazon Basin." *Garden* 6 (1) (1982): 1-36.

Gomez-Pompa, A., Vazquez-Yanes, C. and S. Guevera. "The Tropical Rain Forest: A Nonrenewable Resource." *Science* 177 (1972): 762-765.

Higgins, G.M., Kassam, A.H., Naiken, L., and M.M. Shah. "Africa's Agricultural Potential." *Ceres* 14 (5) (1981): 13-26.

Hrdy, S.B. and W. Bennett. "The Fig Connection." *Harvard Magazine* 82 (1) (1979): 24-30.

Hughes, J.D. and J.V. Thirgood. "Deforestation in Ancient Greece and Rome: A Cause of Collapse?" *The Ecologist* 12 (5) (1982): 196-208.

Iltis, H.H. "From Teosinte to Maize: The Catastrophic Sexual Transmutation." *Science* 222 (4626) (1983): 886-894.

Janzen, D.H. "The Deflowering of Central America." *Natural History* 83 (4) (1974): 48-53.

Jordan, C.F. "Amazon Rain Forests." *American Scientist* 70 (4) (1982): 394-402.

Khoshoo, T.N. "Energy From Plants: Problems and Prospects." *Proceedings of the Sixty-Ninth Session of the Indian Science Congress, Mysore, Part II: Presidential Address.* (1982): Lucknow, India.

La Bastille, A. "How Menacing is Acid Rain?" *National Geographic* 160 (5) (1981): 652-681.

Lovejoy, T. "Fading Tropical Forests." *Defenders* 56 (2) (1983): 2-5.

McMahan, L. "Cynthia Giddy's Nursery for Cycads." *Garden* 8 (4) (1984): 6-7, 32.

Meijer, W. "Endangered Plant Life." *Biological Conservation* 5 (3) (1973): 163-167.

Nations, J.D. and D.I. Komer. "Rainforests and the Hamburger Society." *Environment* 25 (1983): 12-20.

Oldfield, M.L. "Tropical Deforestation and Genetic Resources Conservation." *Studies in Third World Societies* 14 (1981): 277-345.

Pictorial China. "Chinese Gardens (18)", "Nature Preserves(19) & (20)." *China Pictorial, Beijing, China* (1983).

Poore, D. "The Value of Tropical Moist Forest Ecosystems and the Environmental Consequences of their Removal." *Unasylva* 28 (112-113) (1976): 127-146.

Raven, P.H. "The Importance of Preserving Species." *Fremontia* 11 (1) (1983): 9-12.

Risser, J. "A Renewed Threat of Soil Erosion." *Smithsonian* 11 (12) (1981): 120-131.

Rubinoff, I. and N. Smythe. "A Jungle Kept For Study." *New Scientist* 95 (1319) (1982): 495-499.

Thorsell, J.W. and G. Wood. "Dominica's Morne Trois Pitons National Park." *Nature Canada* 5 (4) (1976): 14-16, 33-34.

Tomaselli, R. "The Degradation of the Mediterranean Maquis." *Ambio* 6 (6) (1977): 356-362.

Train, R. "Biological Diversity: The Ecological Basis for Sustainable Development." *World Resources Institute Journal* '84 (1984): 27-33.

Wang, S. and J.B. Huffman. "Botanochemicals: Supplements to Petrochemicals." *Economic Botany* 35 (4) (1981): 369-382.

White, P.T. "Tropical Rain Forests: Nature's Dwindling Treasures." *National Geographic* 163 (1) (1983): 2-47.

Index

251

Picture Credits

courtesy Istanbul University; 181 National Museum of American Art, SI, Transfer From Museum of Modern Art; 182 Reprinted from *Caribbean Review*, Florida International University, Miami, FL 33199, Vol. XII, No. 3; 183 *Voyages dans L'Amérique du Sud*, Page 362, Photo by Ed Castle; 184 Kjell B. Sandved; 185 Collection of Business Americana, SI; 186 Courtesy of Eli Lilly and Company; 187 (top) Edward S. Ayensu; (bottom) *Lilly World News*, Courtesy of Eli Lilly and Company, Photo by Ed Castle; 188 (top) Edward S. Ayensu; (bottom) Christopher Grey-Wilson; 189 (top) Edward S. Ayensu; (bottom) Joe Goulait/SI; 190-192 Edward S. Ayensu; 193 (top) Edward S. Ayensu; (bottom) Kjell B. Sandved; 194-199 Edward S. Ayensu.

Part IV: pp. 200-201 Thomas R. Soderstrom; 203 ©Earth Satellite Corporation, under the GEOPIC trademark; 204 Edward S. Ayensu; 205 (top) Robert W. Read, Department of Botany/SI; (bottom) Agency for International Development; 206 (top) Ann Hawthorne; (bottom) Dr. Philip Ross, National Academy of Sciences; 207 NASA; 208 Soil Conservation Service/USDA; 209 Kjell B. Sandved; 210 (left) *The Gardeners' Chronicle*, July 15, 1882, Photo by Vic Krantz; (right) *Curtis's Botanical Magazine*, 1871, Plate 5904, Photo by Joe Goulait/SI; 211 *Flora Brasiliensis*, Plate 35, Photo by Joe Goulait/SI; 212 (left) Kjell B. Sandved; (right) David B. Lellinger, Department of Botany/SI; 213 Department of Botany/SI; 214 Donald R. Davis, Department of Entomology/SI; 215 (left) Courtesy of the Trustees of the British Museum, London; (right) Christopher Grey-Wilson; 216 Agency for International Development; 217 Photograph by Eliot Elisofon, National Museum of African Art, Eliot Elisofon Archives, SI; 218-219 Stephen L. Buchmann; 219 (top) Noel Vietmeyer; (right) Dr. Allen C. Gathman, Plant Sciences Department, University of Arizona; (bottom) Stephen L. Buchmann; 220-221 Noel Vietmeyer; 223 (top) Noel Vietmeyer; (bottom) Edward S. Ayensu; 224 Joe Goulait/SI; 224-225 Noel Vietmeyer; 227 Inter-American Development Bank; 228 Noel Vietmeyer; 229 Agency for International Development; 230 (top) World Bank Photo by Ed Huffman; (bottom) Edward S. Ayensu; 231 (top) Joe Goulait/SI; (bottom) Edward S. Ayensu; 232-233 Edward S. Ayensu; 234-235 Valerie Brown.

Part V: pp. 236-237 Agency for International Development; 238 Anni-Siranne Coplan; 239-241 Carl Purcell/Agency for International Development.

Back Matter: p. 242 Kjell B. Sandved; 256 Photograph by Eliot Elisofon, National Museum of African Art, Eliot Elisofon Archives, SI.

Jacket: *Nymphaea* water lily, Java; Fern *(Dipteris conjugata)*, Mt. Kinabalu, Sabah;
Page 1: Flowering hepatica, Michigan, United States;
  2-3: Succulent *Dudleya traskiae*;
  4-5: Paramo vegetation, Colombia;
  6-7: Close-up of leaf structure;
  8-9: Left to right: *Pteris* sp.; liana-tendrils; *Epidendrum ciliare*, epiphytic orchid; Spanish moss, an air plant; liana-tendrils curled; *Spathicarpa sagittifolia*;
   10: Cardinal flower of Costa Rica, *Rechsteineria cardinalis*.

Dylan Thomas, *Poems of Dylan Thomas*. Copyright 1939; New Directions Publishing Corporation. Reprinted; permission of New Directions. See page 27.

Our Green and Living World was designed by Carol Hare Beehler, Smithsonian Institution Press. Typeset in Aster by VIP Systems, Inc., Alexandria, Virginia. Printed on Chromosatin 120 gsm by Balding + Mansell, Wisbech, Cambs. Bound with Colorplan Omar Brown endpapers and Joanna Devon Brown cloth. Editorial Production by the Smithsonian Institution Press, Washington, D.C. Published in September, 1984, on the occasion of the Plants Campaign of the World Wildlife Fund and the International Union for Conservation of Nature and Natural Resources, (Year of the Plant), and the 400th Anniversary of the Cambridge University Press.